A Field Guide to Climate Change

A FIELD GUIDE TO
CLIMATE
CHANGE

Understanding the Problems

Adam Briggle

broadview press

BROADVIEW PRESS – www.broadviewpress.com
Peterborough, Ontario, Canada

Founded in 1985, Broadview Press is a fully independent academic publishing house owned by approximately twenty-five shareholders—almost all of whom are either Broadview employees or Broadview authors. Broadview is supported by a collaboration with Trent University, a liberal arts university located in Peterborough, Ontario—the city where Broadview was founded and continues to operate. Broadview is committed to environmentally responsible publishing and fair business practices.

Library and Archives Canada Cataloguing in Publication

Title: A field guide to climate change : understanding the problems / Adam Briggle.
Names: Briggle, Adam, author.
Description: Includes bibliographical references and index.
Identifiers: Canadiana (print) 20240355423 | Canadiana (ebook) 20240355555 | ISBN 9781554815937 (softcover) | ISBN 9781770489592 (PDF) | ISBN 9781460408834 (EPUB)
Subjects: LCSH: Climatic changes. | LCSH: Climate change mitigation.
Classification: LCC QC903 .B75 2024 | DDC 363.7—dc23

Broadview Press handles its own distribution in Canada and the United States:
PO Box 1243, Peterborough, Ontario K9J 7H5, Canada
555 Riverwalk Parkway, Tonawanda, NY 14150, USA
Tel: (705) 482–5915
email: customerservice@broadviewpress.com

For all territories outside of Canada and the United States, distribution is handled by Eurospan Group.

Broadview Press acknowledges the financial support of the Government of Canada for our publishing activities.

Edited by Robert M. Martin
Book Design by Em Dash Design

Broadview Press® is the registered trademark of Broadview Press Inc.

PRINTED IN CANADA

For Max and Lulu—in hopes that your generation learns how to be Earthlings.

Contents

Preface

Writing in the wake of the first atomic bomb explosions, the thinker Günther Anders (1962) argued that we have become "inverted utopians." Traditional utopians are unable to produce what they can imagine. By contrast, we are unable to imagine what we are producing. We have built a world that we cannot fathom, but we have to try, because with great power comes great responsibility. Our task is to "think what we are doing" (Arendt 1958, x).

Our growing powers create an ever more urgent moral imperative not just to think, but to cooperate. Climate change is perhaps the most consequential example. It is a tangle of problems that will define twenty-first-century global civilization. The impacts of climate change are here and are getting worse. But many solutions are also here and are getting better. With care, clear thinking, and solidarity, we can act in hope of creating a thriving future.

Only recently have schools started listening to students' pleas for climate education. Most of my college students never learned about climate change in school. They tell me that it feels like a series on Netflix: they've heard about it, but they don't understand the plot or the main characters. This leaves them feeling anxious and confused in the swirl of polarized politics and click-bait media. There are so many narratives about climate change out there. Which ones are reasonable and which ones are not? How do we make sense of all this!?

This book aims to help. It offers tools for climate literacy, which is about more than science and engineering. Climate change is foremost a political phenomenon. It is about power, money, meaning, and values. Climate change respects no borders, reminding us that to be is to be with others. To be climate literate is to understand how problems get framed and contested in a multispecies and multicultural plurality. That's why this book takes a problem-posing approach to climate change. It offers tools to focus attention and organize thought and action.

I have worked on this book over the past five years of teaching about the human condition in the age of science and technology. I am grateful for my students—through loving struggles, they taught me what climate change means to them and how to teach and learn about it. I am in debt to my mentors Fr. Mark Thamert, Carl Mitcham, Robert Frodeman, Steve Fuller, and Roger Pielke, Jr. A special thanks goes to Maggie Brown and my wife, Amber, who both helped with the thinking in these pages. Finally, I am grateful for the guidance of Robert Martin, Stephen Latta, Archie Fields, and for all the good people at Broadview Press.

References

Anders, Günther. 1962. "Theses for the Atomic Age." *The Massachusetts Review* 3 (3): 493–505.

Arendt, Hannah. 1958. *The Human Condition.* Chicago: University of Chicago Press.

Teaching and Learning in the Tangle

Global Annual CO₂ Emissions

Carbon dioxide (CO₂) emissions from fossil fuels and industry.[1] Land use change is not included.

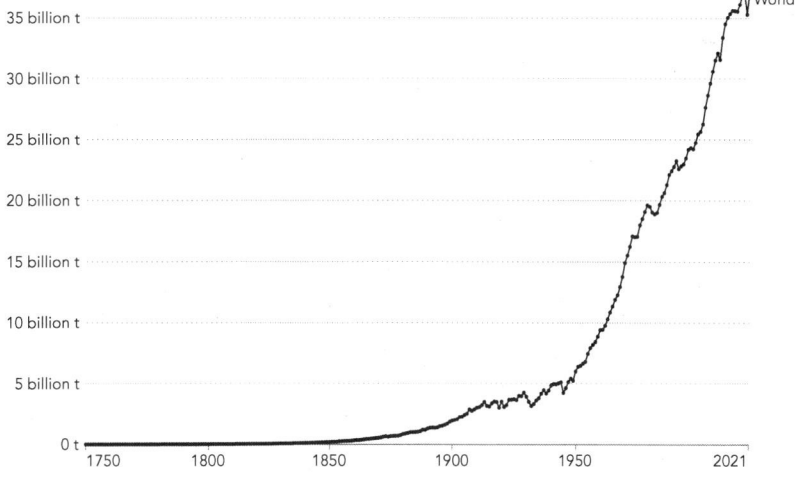

1. Fossil emissions: Fossil emissions measure the quantity of carbon dioxide (CO₂) emitted from the burning of fossil fuels, and directly from industrial processes such as cement and steel production. Fossil CO₂ includes emissions from coal, oil, gas, flaring, cement, steel, and other industrial processes. Fossil emissions do not include land use change, deforestation, soils, or vegetation.

FIGURE 0.1

Picture in your mind a barrel of oil. That's a 42-gallon drum. Then imagine 1,100 of those barrels. Now count to one. It's all gone. Every second, the global economy combusts that much oil. But that's not all. Picture a stack of coal two feet long, two feet wide, and 10 feet high. Now multiply that by 270. That's 270 tons of coal, and it too is combusted *every second*. There's more. Imagine 45 Olympic-size swimming pools, which is

11

about 4,000,000 cubic feet. That's equivalent to the amount of natural gas burned every second.[1]

The figure above shows the carbon dioxide emissions from all that fossil fuel combustion. It is one way to represent a massive energy flow. Indeed, we have become geologic agents of change, shaping the entire planet through agriculture and by moving enormous quantities of stuff from the lithosphere[2] into the atmosphere and oceans.

Energy is change. This is a book about change. To appreciate that, we have to see ourselves as always in the process of becoming. Changes are shaping us and we can learn to shape the changes. To me, this is what problem-posing education is all about. It is a process of inquiry to spark critical consciousness of the world we have been thrown into, how it is changing, and how we might participate in those changes (see Freire 1970). It's less about memorizing facts than gaining the skills to see how facts are created, contested, and mobilized in the pursuit of values.

Problem-posing helps us to question the world and to ask the ultimate question: what should we do? That word "we" is key. My students ask what they can do as individuals to address climate change. Personal choices are important, but they won't usher in the structural changes required. Indeed, obsessing over your personal "carbon footprint" can be a distraction from the transformative changes to the systems that shape and constrain our choices as individuals. We need collective action; we need to act as citizens (not just consumers) to push together on the parts of the system where meaningful change can happen. I want you to see yourself as part of something bigger so that you can find places to contribute your gifts.

Your Mental Model

Climate literacy entails building our knowledge base, but also recognizing the impossibility of knowing it all. There is so much information and even more uncertainty. Studying climate change can feel like drinking from a firehose. The human mind filters this complex flood of information through conceptual lenses to form a mental model, a simplified picture of reality through which we interpret experience. For many people, their mental

---

1   Figures from Energy Worldometer: https://www.worldometers.info/energy/.
2   The outermost crust of the Earth, approximately 60 miles deep.

model is shaped by their religious beliefs. This makes it essential to talk about religion as a part of climate politics, and to do so in a way that focuses on shared values (Hayhoe 2021).

Mental models include our core beliefs and values; these result in the opinions that we hold. There are many ways to talk about this notion of a mental model. For example, the writer Nadia Asparhouhova (2022) argues that there are seven "tribes" of climate change. The Yale Program on Climate Change Communication groups mental models into what they call the "Six Americas" on climate change beliefs:

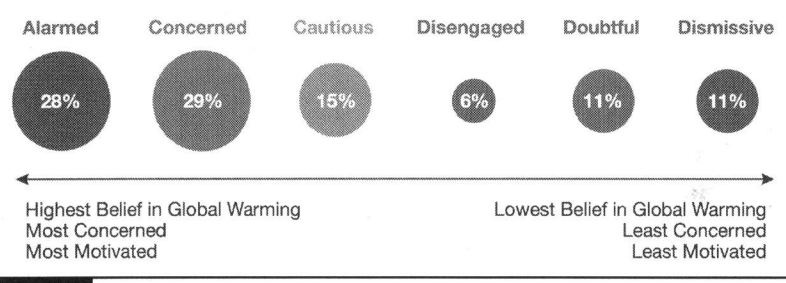

FIGURE 0.2

Mental models are why the best ways to talk about climate change start with our values rather than a list of facts. They explain how, even though you may never have studied climate change, you probably have a kind of feeling or mood about the topic. Climate literacy requires critically examining the way you and others filter things to form different models of the world and the different problem framings that result from them.

Note that the majority of Americans are alarmed (urgent threat now) or concerned (serious threat but not pressing now) and another size-able group is cautious (curious but not sure what to think). Only a small minority is dismissive. In Canada and many other countries, there is even less dismissiveness. The fossil fuel industry and wealthy political groups have convinced us that climate change is more controversial than it really is. As a result, most people who care about climate change don't talk about it that much. This creates a self-reinforcing "spiral of silence," which, like the "bystander effect,"[3] can perpetuate passive climate denial. People feel uneasy

---

3   An onlooker is less likely to help someone in an emergency if there are other bystand-ers around.

but see everyone else going about their lives as if everything is fine (even though they are worried too!). So, if nothing else, it's important to broach the topic. You are likely to find that others are also concerned and eager to talk about it.

Partisan polarization around climate change is less pronounced among younger generations than it is with older folks. Young people generally acknowledge that humans are changing the climate. One of my students who paid for college with money earned welding pipes for the oil and gas industry, told me that he and his friends at work recognize that climate change is real. Having seen the sheer scale of oil refineries in Houston and the gas flares on the Bakken Shale in North Dakota, they can't help but worry. They don't doubt the science, and they think that something must be done. But *what*? And *who* is responsible? And what about their jobs and the families that depend on those paychecks?

This student is in the "concerned" category: he thinks climate change is serious, but it's not necessarily a top priority. He was not doubtful or dismissive. Nearly all the debates I have in my classes are between those who are alarmed about the "climate crisis" and those who are cautious or concerned about "climate change." This mirrors the bigger social picture. Indeed, climate scientists are increasingly having conversations about what language to use, with more and more of them talking and writing about "climate emergency" and not just "climate change."

I also find these different mental models at work among other academics who study climate change. Two of my mentors are an example. Both recognize the seriousness of climate change and the need for action. One is certain that the future is bright. He is confident that we will manage the problems posed by a changing climate. He focuses on evidence showing upward trends in global human development. We just need more high-tech solutions and economic development so that we get smart, wealthy, and resilient enough to solve the problems of our own creation. My other mentor, though, thinks that we are facing a massive, near-term breakdown. He focuses on evidence about the growing human impacts on our planet. If it is inevitable, then we will need "deep adaptation." We will have to relinquish high-tech ways of life, cultivate mutual aid, and restore pre-industrial practices.

Here you can see how different problem-framings can come from different mental models: do we need more or less techno-economic development? These opposing views also show us how political reasoning requires thinking through paradoxes. As my two mentors show, climate change is a paradox—it is the best of times (unprecedented human development and wealth) and the worst of times (unprecedented inequality and peril to eco-systems). Both perspectives contain truth, but how can that be, and are they equally true?

In this book, I treat denialism (dismissiveness) for what it is: a false view that is waning but is still problematic. Outside of that, I have tried to remain as neutral as possible in covering the spectrum of reasonable views on climate change—from caution to alarm. I want to help you understand different perspectives as you critically evaluate your own mental model.

Personally, I have come to embrace a model of cautious optimism. Yes, there are substantial risks facing twenty-first-century civilization. We can't rule out the possibility of a grim future, and the path to achieving our climate goals is daunting. And yet, climate policies are starting to make earlier dire predictions far less likely (see Wallace-Wells 2022). We are making real progress, and there are many good initiatives to which you can add your skills and passions. I'm not endorsing blind optimism—as if tech-nology or the free market ensure that climate change will solve itself. I am talking about the bracing optimism of believing in ourselves—that we can take the reins and solve the enormous problems at hand.

Despair and cynicism are not helpful moods. So, being alarmed is quite reasonable, but don't let it drown you in hopeless pessimism. We are living at an incredible point of inflection in human history, indeed in the multi-species history of Earth. The stakes for the decisions we make now are very high. I think this is exciting and invigorating. History has its eyes on us. Let's rise to the occasion!

Climate Literacy Is More Than Science

Through fossil fuel combustion and changing land uses like agriculture and deforestation, humans are emitting greenhouse gasses like methane (CH_4) and carbon dioxide (CO_2).[4] These gasses trap heat and warm the planet. This is global warming. The planet is already on average 1.2°C warmer now than in pre-industrial times. The Intergovernmental Panel on Climate

---

4   The proportion of GHG emissions is roughly 74 percent CO_2, 17 percent CH_4, 6 percent Nitrous Oxide (N_2O), and 3 percent F-gases like HFCs (Hydrofluorocarbons) and CFCs (Chlorofluorocarbons).

Change (IPCC) is the world's authority on climate science. In a 2021 report, the IPCC wrote, "It is unequivocal that human influence has warmed the atmosphere, ocean and land" (4).

This warming, in turn, is altering weather patterns and extremes in ways that cause harm. This is climate change. All that extra heat in the system—like boiling water in a pot—is causing things to stir around. Pathogens, animals, soils, plants, humans, and entire ecosystems are on the move. Extreme weather events are happening with more intensity and greater frequency. Our most likely future may not be total collapse, but it also won't be status quo. We are in for transformative changes.

In short, climate change is here and it is impacting the natural and social systems that supply our food, water, energy, and more. We are changing the conditions on Earth that have allowed civilization to flourish. The impacts are primarily caused by the world's wealthy and they hit the world's poor the hardest. This is a profound injustice that we have barely begun to reckon with.

Although science has alerted us to this situation, it cannot tell us what to do. The IPCC said this in a 2001 report:

> Natural, technical, and social sciences can provide essential information and evidence needed for decisions on what constitutes "dangerous anthropogenic interference" with the climate system. *At the same time, such decisions are value judgments* determined through socio-political processes, taking into account considerations such as development, equity, and sustainability, as well as uncertainties and risk. (2, emphasis added)

As the geographer Mike Hulme (2020) notes, "the fact that humans are altering the world's climate is absolutely clear, the significance of this fact is not self-evident" (3). What does it mean, and who should do what about it? In other words, we need to think about values in a tangle with the sciences. This is what problem-posing is all about.

And this explains why schools have done a poor job on climate literacy. It's not that teachers don't care. In a national survey, 86 percent of American teachers said that they believe climate change should be taught. Yet only 45 percent of them teach it (Newall and Patino 2019). The main reasons given are: "it's not related to the subject I teach" and "I don't know enough about it." We don't teach about climate change, in other words, because it is not a subject. It is a problem—a tangle of problems. Climate change does not fit into the boxes that organize education. The disciplinary model tells teachers that they have to be subject-area masters. Yet no one can be an expert in something as complex as climate change. So, it falls through the cracks. Another very important reason for this reluctance is that when curricula are strongly controlled by school boards or parents, teaching anything about climate change may get teachers in trouble with their overseers.

When we do teach about climate change, we tend to box it in as a strictly scientific topic. It's understandable. After all, this allows us to cram things into educational categories. Plus, "the climate" is experienced through the mediation of scientific knowledge and instruments. And a focus on science allows teachers to dodge touchy political and ethical debates.

Of course climate science is important. And, sadly, we often do a poor job on this front. One report found that many teachers give equal time to perspectives that raise doubt that humans are causing climate change (Plutzer et al. 2016). This is irresponsible. It is an example of *the first pitfall: false equivalence.* Not all statements are equally true, not all arguments are equally strong, and not all sources of information are equally credible. There is often room for interpretation and legitimate disagreement, but reality and the truth are not infinitely flexible things.

There are wealthy corporations, special interests, and politicians who obstruct climate action in bad faith (Oreskes and Conway 2010). Take, for example, the campaign of climate denial waged by ExxonMobil. More recently, outright denial is less of a problem than "greenwashing" attempts where the rich and powerful might appear to take climate change seriously but are just setting up a smokescreen to hide business as usual. ExxonMobil lobbyists, for example, admitted that they supported a carbon tax in the US only because they knew it was a political non-starter. It was just a PR stunt to delay action (McGreal 2021). Climate pledges by nation states and corporations are part of the confusing rhetoric swirling around climate change. Sometimes, they are nothing more than delay tactics.

So, part of climate literacy is getting clear about where there is no longer any legitimate scientific controversy, when a claim is a lie or at best a half-truth, and when a promise is nothing more than hot air designed to prolong the status quo that benefits the powerful and wealthy.

There is another pitfall to worry about, though. Seeking to avoid false equivalence, we might start insisting that all we have to do is follow the facts or listen to the experts. Yet climate change is full of uncertainties and disputes even among experts. In trying to reduce all this complex, nuanced ambiguity

Global Warming's Six Americas

FIGURE 0.3

down to a simple story of "facts vs. fiction," we risk the second pitfall: *false binary*. It's not usually a matter of truth versus lies or the good guys versus the bad guys. We won't resolve disagreements or find the best course of action by administering a simple litmus test of belief or denial (see Hulme 2009). We have to know *which* truths or uncertainties to highlight and how to balance different benefits, costs, and risks in search of common goods.

The climate debate has largely shifted from "is there a problem?" to "what should we do about it?" Much of climate politics is legitimate disagreements about values trade-offs and how to interpret the situation (this is how different mental models often clash). To insist that we just have to "follow the science" perpetuates the information deficit myth: more and better science will pave the way to better and easier policies. It does not work this way. So, the trick to climate literacy is knowing how to strike this dynamic balance of both insisting on facts and acknowledging that "the facts" won't speak for themselves.

What Does Climate Change Look Like?

That humans are altering Earth's climate is an unequivocal fact. That we should do something about it is the consensus of the international community. We need to rapidly *reduce* greenhouse gas emissions, get more *resilient* in a world that is already changing, and *pay* for all of this. Thus, we have the three main categories of climate action: *mitigation*, *adaptation*, and *finance*.

This seems straightforward. So, why is climate politics so difficult? As the journalist Noah Gallagher Shannon (2022) put it, "While the math of decarbonization and electric mobilization is clear, the future lifestyle it implies isn't always." It takes an imaginative leap into unknown ways of being in the future. Consider eight real-life examples to get a practical sense of the challenging tangles involved.

* A city in Texas could require new homes to use zero-carbon materials and energy sources. It would save money and reduce emissions in the long run, but it would increase the costs of new homes in the short-term. Housing is scarce, so prices are high for working-class families. What's the right thing to do? Who should decide that?

* Climate change has brought increased flooding to parts of Louisiana. Maybe some people should be relocated. But who is going to pay, and what if folks don't want to move? Should we build giant seawalls to protect against flooding? Who would pay for those?

* Wind and solar power are expanding as they get cheaper, yet they are intermittent sources of electricity. How do we maintain reliable electricity? Batteries? What kind? And what are their environmental and social impacts? What about nuclear power or hydrogen or geothermal?

* We might be able to cool the planet by injecting sulfates into the stratosphere. Who should have the authority to decide whether we run a planetary experiment like that? Or, we could pull more CO_2 out of the atmosphere and inject it back underground. But that is very expensive. Should we fund these carbon capture technologies ... even if they benefit the oil and gas industry?

* There is an enormous amount of carbon stored in the peatlands of the Democratic Republic of Congo (DRC). Keeping that carbon in the ground is essential to achieving climate goals. But draining the peatlands for agriculture, which would release the carbon into the atmosphere, would boost local development. So, what should be done about the peatlands (a ticking "carbon bomb") in this part of Africa? Should developed nations pay the Democratic Republic of

Congo to keep the carbon in the ground? Would that money go to the people who need it the most?

* Kiribati is an island nation in the Pacific Ocean that is threatened by rising sea levels and changing storm patterns. The citizens of Kiribati are becoming climate refugees as their home disappears. Where can they go? What about their culture? Who has a responsibility to help them? Similar questions pertain to Indigenous peoples in Arctic regions where sea ice is melting. How can we prevent a future of desperate climate refugees, hardened borders, hoarding, and conflict?

* Decarbonizing the transportation sector means lots of electric cars, many of which require lithium for their batteries. A large portion of lithium supplies come from mines in the ecologically sensitive salt flats of Chile. And most cobalt (another key mineral for batteries) comes from the DRC, where labor conditions in the mines are atrocious. How should we balance local environmental and human rights concerns with concerns about climate change? Does solving global climate change require creating local sacrifice zones? How should that be decided?

* Warming temperatures have allowed beetles to devastate forests in British Columbia. This has led to a lumber shortage for home building in the US. One proposal is to open up more logging in lands that are currently protected. Some say it would be good to thin out those forests, others say we need to stop cutting trees. What should we do?

Notice how climate change doesn't appear in human affairs as a stand-alone issue. It is tangled up with everything: economics, law, science, engineering, and art. It shows up in land-use and energy policy, immigration, building codes, healthcare, transportation, geopolitics, and more. This is why significant climate legislation might be called something that sounds unrelated like the "Inflation Reduction Act." It's also why we sometimes succeed on climate goals by *not* talking about climate at all—maybe air quality, job creation, or energy security are better frames.

These stories show again that the challenge is not just acknowledging the climate science (avoiding false equivalence), but also doing the moral and political work of problem-solving (avoiding the false binary conclusion that we just have to listen to the facts). Imagine the complexities entailed in any one of those situations above. As you lead your lives, climate change will follow you. It will appear in your jobs and your personal and civic lives. You will need specific information for those contexts, whatever they may be. Yet the general tools in this book will help as you think through questions like:

* What actions are needed to thwart the risks of climate change? What are their costs and benefits? Are prices reflecting the actual costs? Who will bear the costs? Are they distributed fairly?
* What are our responsibilities? How much should we consider future generations and non-human species? What do we owe to people on the other side of the planet?
* Who should have what kind of power in decision-making? How do we forge alliances across differences in the pursuit of common interests? Which compromises are acceptable and which are not? What is political will and how do we manifest it?
* How are problems being framed and which framing is best? How should we act when things are uncertain? What are the unintended consequences of proposed solutions?
* What is the proper reach of government? What is the proper role of private capital and free market competition? What constitutes effective activism?

On Values

It is okay if you don't know the answers to these questions. Neither do I! Teaching and learning in the climate tangle is not as much about acquiring knowledge (though that is part of it) as it is about asking questions and gaining insights. Like Socrates, we are not trying to be experts. We are learning how to thoughtfully travel through the relevant parts of the tangle.

Values are the strands we will follow through the tangle. They are the goals that determine how we see the climate problem. Values are what matter—our ideals and aspirations: Freedom, justice, prosperity, happiness, health, beauty, pleasure, knowledge, power, and more. They filter that flood of information, shaping our mental models. Sometimes we share values.

Other times we disagree. We might value the same word, like "sustainability," but mean different things by it. Values are interpretive but that does not mean they are subjective and irrational. We can reason together and get more clarity; we might even find common ground.

 **THERE ARE VALUES IN SCIENCE.** Epistemic[5] values are the norms governing how to pursue knowledge, what to count as knowledge, and how to communicate uncertainties. There are ethical values about, for example, what to study, how to do it, and the responsibilities of scientists. There are questions about whether we all have a duty to learn and, once we know, to act.

 **THERE ARE VALUES IN MACHINES.** Technologies embody values. They are the result of choices about safety, beauty, money, utility, sustainability, and more. So much of climate change is about *instrumental values*:[6] what technologies should we fund, build, and deploy and by what rules should they be governed? What values do and do not get built into machines? Whose visions get built? Why theirs?

 **THERE ARE VALUES IN POLITICS.** Human action is not an algorithm. We don't plug in information and spit out a decision. Humans are the "political animal" because we guide our actions through stories, words, and persuasion. Politics is about human plurality—we are all unique, yet we can only thrive together. We are not a hive-mind with a single consciousness. Thus, politics is about power differentials: who gets what, when, how, and why? Political reasoning entails seeing how different factions frame problems. The value of freedom is particularly important in climate politics. There are many freedoms that often clash: freedom to consume and innovate, freedom from government, freedom from climate disasters, freedom to inhabit a healthy environment, etc.

---

5   Having to do with knowledge.

6   Something is of instrumental value if it results in something else of value. Something has intrinsic value, by contrast, insofar as it is valued in itself.

Media Literacy and Critical Thinking

Climate change is real, a matter of change in average climate over time, so never directly experienced, but always mediated by the variety of methods and concepts for discovering it. This means there are many ways to understand it, and many lenses through which to view it. Diverse ways of knowing and thinking can be a strength, but recall the pitfall of false equivalence: not every view is equally accurate and not every claim is equally true.

We face the timeless philosophical question: How can we conduct our reasoning to avoid being deceived? The mediated nature of climate change makes this even trickier, because rarely can we test claims using sensory experiences. Information and arguments get to us through various media: scientific reports, government agencies, non-governmental organizations, corporations, advertisers, mainstream media, social media, and even AI bots. Critical thinking, then, means paying attention to the credibility, reliability, and rhetorical tactics of your sources.

Climate literacy requires critical thinking skills to spot logical fallacies and determine which arguments are sound. This is hard to do, but we can start by laying out the kinds of skills required: interpretive, verification, and reasoning (see Hughes and Lavery 2008):

1. *Interpretation*. Human languages are not as clear cut as binary code. Communication is often full of vagueness, irrelevancies, and ambiguity. Interpretive skills help us to understand the meaning of statements and clarify the values and assumptions being used to frame problems. What is included and what is excluded?

2. *Verification*. Even when we understand the meaning of a statement, it may be hard to determine whether it is accurate. Verification skills help us to evaluate the truth value[7] of statements. The technical complexities of climate change make verification difficult. We cannot possibly know enough to test all statements on our own, so whom can we *trust* and under what conditions are we justified in forming beliefs on the basis of trust?

---

7   The truth values are *true* and *false*. A straightforward statement has one of these as its truth value.

3. *Reasoning*. Truth is a property of statements; it is what we are trying to verify. Logical strength, by contrast, is a property of inferences, which are about the relationship between statements. Inferences are how we build arguments by moving from one thought to the next with words like "since," "thus," "because," or "it follows that." Reasoning skills are about assessing the strength or soundness of arguments.

We'll work on all skills throughout the book. But the emphasis of Part One (Big Picture and Fundamentals) is on interpretation, Part Two (Climate Sciences) focuses on verification, and Part Three (Politics, Ethics, and Policy) is primarily about reasoning as we examine the ways science, politics, and ethics get tangled into arguments about what we should do.

The New Greatest Generation

The people who fought in World War II and contributed on the home front are called "the greatest generation." Through hardship and sacrifice, they defended liberty and human dignity against totalitarianism. They risked their lives for these ideals.

The post-war years brought an economic boom. Average global wealth per capita went from about $3,000 then to $15,000 now. Of course, that wealth is not equitably distributed, but the share of the world's population living in extreme poverty has dropped significantly and life expectancy has risen in much of the world. At the end of the war, electrification had only just begun, and now over 90 percent of the world's population has access to electricity.

Climate change is, in part, the unintended consequence of this development project. People often liken climate change to the menace of fascism in World War II. Like the allied response to the axis powers, our response to climate change will require economic and social mobilization, political will, personal courage, and technological development. It is less clear whether it will require voluntary hardship—most people are not willing to sacrifice conveniences for climate action.

Indeed, the war analogy only goes so far. We are not fighting a foreign enemy. If anything, we are fighting our own way of life. Climate change is the unintended bad consequences of lots of good things like electricity, heat, and clean water. This tangling of good and bad is the trickiest knot of all.

If we were to bomb coal plants and pipelines, poor and working-class folks would be hit the hardest as energy prices spike. And yet, many of the same vulnerable groups that would be harmed by rapid disentanglement from fossil fuels are also most at risk from our current tangle. Communities of color, for example, are disproportionately impacted by many of the polluting facilities that power the modern world. And in many places around the world, women are most vulnerable to climate impacts.

We can't fight yesterday's wars. Eighty years ago, the human footprint was relatively small. Now, we are a planetary force. The greatness of that former generation consisted in the steadfast defense of noble ideals. The greatness of this new generation will consist in the expansion of our moral imagination and responsibilities to match the scope of our technological powers and economic labors. In an age marked by individualism and deep divisions, we have to learn *how to think and act together as Earthlings* on the only planet we've got.

Activities and Questions

1. In "Global Warming's Six Americas," which category best fits you? Where does communication break down with others in different categories? What would it take to get you to shift into another category?

2. Watch the film *Anote's Ark* about Kiribati. What does climate justice look like in this case? Is it even possible? Who has what responsibilities?

3. Watch the film *Merchants of Doubt* about campaigns to delay climate action. How does this film frame the problem? What other framings can you imagine?

4. In 2022, climate activists with the group Just Stop Oil threw tomato soup on a Van Gogh painting in London's National Gallery of Art. Research this and related actions. Is this the right thing to do? What makes activism good or bad?

References

Asparouhova, Nadia. 2022. "Mapping Out the Tribes of Climate." https://nadia.xyz/climate-tribes#climate-change-isnt-about-evangelism-anymore.

Freire, Paulo. 1970. *Pedagogy of the Oppressed*. New York: Continuum International.

Hayhoe, Katherine. 2021. *Saving Us: A Climate Scientist's Case for Hope and Healing in a Divided World*. New York: Simon & Schuster.

Hughes, William, and Jonathan Lavery. 2008. *Critical Thinking: An Introduction to the Basic Skills*. Peterborough, ON: Broadview Press.

Hulme, Mike. 2020. "Is It Too Late (to Stop Dangerous Climate Change)?" *WIREs Climate Change* 11, no. 1 (January/February): 1–7.

—. 2009. *Why We Disagree about Climate Change: Understanding Controversy, Inaction and Opportunity*. Cambridge: Cambridge University Press.

IPCC. 2001. *Climate Change 2001: Synthesis Report. Contributions of Working Groups I, II, and III to the Third Assessment Report*. Cambridge: Cambridge University Press.

—. 2021. *Climate Change 2021: The Physical Science Basis. Contribution of Working Group I to the Sixth Assessment Report*. Cambridge: Cambridge University Press.

McGreal, Chris. 2021. "ExxonMobil Lobbyists Filmed Saying Oil Giant's Support for Carbon Tax a PR Ploy." *The Guardian*, June 30.

Newall, Mallory, and Thomas Patino. 2019. "Teachers Agree that Climate Change Is Real and Should Be Taught in Schools." Ipsos News and Polls, April 22. https://www.ipsos.com/en-us/news-polls/teachers-agree-climate-change-real-and-should-be-taught-schools-04-22-2019#:.

Oreskes, Naomi, and Erik Conway. 2010. *Merchants of Doubt: How a Handful of Scientists Obscured the Truth on Issues from Tobacco Smoke to Global Warming*. New York: Bloomsbury.

Plutzer, Eric, et al. 2016. *Mixed Messages: How Climate Is Taught in America's Schools*. Oakland, CA: National Center for Science Education.

Shannon, Noah Gallagher. 2022. "What Does Sustainable Living Look Like? Maybe Like Uruguay." *New York Times*, October 5. https://www.nytimes.com/2022/10/05/magazine/uruguay-renewable-energy.html.

Wallace-Wells, David. 2022. "Beyond Catastrophe: A New Climate Reality Is Coming into View." *New York Times*, October 26. https://www.nytimes.com/interactive/2022/10/26/magazine/climate-change-warming-world.html.

PART 1

The Big Picture and Fundamentals

CHAPTER 1

The Anthropocene and Development

A Ball-and-Cup Depiction of the Earth System

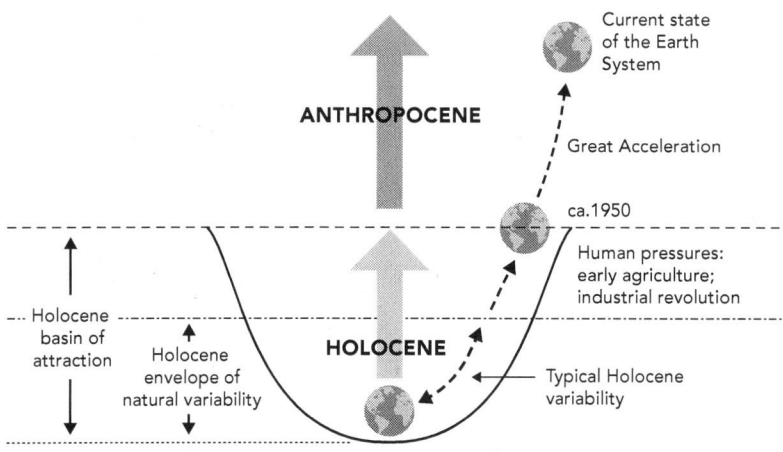

FIGURE 1.1

To begin, let's zoom way out to get our bearings and ask the basic questions: where and when are we? Then we can ask: who and why are "we"? The first set of questions (where and when) *describes* our situation as the Anthropocene or the age of humans (*anthropos* is the Greek term for human). The second set of questions (who and why) *diagnoses and evaluates* our situation: what does the Anthropocene really mean, what ideas and values are behind it, and is it—or could it become—a good thing?

We can also ask how we got here. The processes that have created the Anthropocene are typically called *economic development*. This is all about production and consumption and the infrastructure that supports it (research, roads, ports, electricity grids, water systems, etc.). The Intergovernmental Panel on Climate Change defines development as efforts "to improve human well-being." That term "well-being" is vital and will recur throughout our discussions. What is a good life (or human flourishing) and what social and ecological conditions are necessary for it?

We'll see in later chapters that the international community frames the climate problem in terms of "sustainable development," so it makes sense to begin with this wider context. It helps us to see climate change as just one part of our complex natural-political situation. And it introduces themes about identity, justice, and progress that shape all of climate politics.

Where and When Are We?

On February 14, 1990, the *Voyager 1* space probe was leaving the solar system on its way to deep space. At the request of the astronomer Carl Sagan, NASA had *Voyager 1* take one last picture of Earth from a distance of 3.7 billion miles away. From that distance, Sagan marveled in 1994, Earth looked like a "pale blue dot," a "lonely speck in the great enveloping cosmic dark." It is a small world, after all: you could fit 1.3 million Earths into the Sun.

Every human being has lived out their lives on that "mote of dust suspended in a sunbeam." It makes it seem utterly foolish, Sagan wrote, that people should wage wars to be the temporary masters of a fraction of this little pixel. He said that this image of Earth "underscores our responsibility to deal more kindly with one another, and to preserve and cherish the pale blue dot, the only home we've ever known."

We might visit other planets, but the prospects of terraforming[1] or settling them remain remote. The average surface temperature on Mars, for example, is negative 63°C, the atmosphere is toxic and has only trace amounts of oxygen, there is very little water, we would need pressure suits to survive, there are high levels of UV rays, and the soil is toxic. This is not to mention how space travel ravages the human body.

This brings home a profound truth: humans are adapted to life on Earth. We can breathe, drink, and eat here. Our planet is habitable. We can exist here as inhabitants in habitats where we form our diverse range of habits, that is, our cultures (Rozzi and Poole 2020). More precisely, we are adapted to the current climatic conditions on Earth—the conditions under which our present way of life has developed. The problem is that we are changing those conditions. Figure 1.1 illustrates the idea of tipping from one geolog-

---

1   To transform alien conditions (on another planet, etc.) to be more like those on Earth.

ical climatic regime (the Holocene) into another (the Anthropocene). We have become so powerful as to write a new chapter in Earth's history.

Yet we are not adapted for this new age that we have created. On a deep level, climate change is a physiological problem. Human bodies are adapted to the thermal conditions of the Pleistocene and Holocene, a time period encompassing the past 2.5 million years. This was a time of relative cold for the planet with lots of ice at the poles and widespread, massive glaciers. Our bodies evolved to regulate heat in these conditions. Now, however, CO_2 levels in the atmosphere are as high as they were five million years ago during the Miocene epoch when sea levels were 50 to 100 feet higher and the temperature was much hotter—often beyond the thermal limits that we can endure.

We are not yet at those temperatures, because there is a time lag: it takes a while for the climate system to equilibrate with the amount of CO_2 in the atmosphere. As a philosopher and scientist put it, "The planet will eventually catch up. A Miocene climate is coming. It's like seeing the flash of a far-off explosion, but the shockwave hasn't reached us yet" (Frodeman and Bullock 2022).

Earth is 4.5 billion years old. To make sense of that much time, think in terms of the "Earth Calendar." Condense the entire geologic history of Earth into one calendar year. January 1 represents the formation of Earth and our current moment in time would be December 31, just at the stroke of midnight. Sketching a few highlights of this calendar would look something like this:

The Earth Calendar

JANUARY Water forms	FEBRUARY Single-celled organisms	MARCH Photosynthesis begins	APRIL Early glaciation	MAY Iron and sulfur oxidized	JUNE Great Oxidation Event
JULY Glaciation leads to mass extinction	AUGUST Organisms with nuclei	SEPTEMBER Multi-cellular life	OCTOBER Sexual reproduction	NOVEMBER Cambrian explosion, plants and animals	DECEMBER See below

SOME DATES IN DECEMBER	KEY:
Dec. 2: Tetrapods evolve (animals with four limbs)	Earth is 4.5 billion years old, so ...
Dec. 10–24: Age of dinosaurs	Each month = 375 million years
Dec. 25: Primates evolve	Each day = 12.3 million years
Dec. 31, 11:30 pm: Humans evolve	Each hour = 513,000 years
Dec. 31, 11:58 pm: Agriculture begins	Each minute = 8,561 years
Dec. 31, 11:59 pm and 58 seconds: Industrial Age	Each second = 142 years

FIGURE 1.2

Just as the image of the pale blue dot shows us how small we are, the Earth calendar makes us appreciate how recently we arrived on this little speck. Even dinosaurs, which symbolize ancient life, didn't emerge until mid-December. Humans didn't evolve until the last half hour on the last day. And all the habits of our culture that seem so familiar, like big farms feeding big cities full of machines, are just products of the last few seconds of the last minute on the last day.

Scientists have their own way of organizing geologic time into eons, eras, periods, epochs, and ages. Each category marks out a smaller chunk of time, like month, week, day, hour, and minute. Scientists use a variety of methods to date rocks and demarcate one unit of time from another. Often, they will look at fossils and name a portion of geologic time based on the kinds of life that were dominant. The Mesozoic Era, for example, means "middle life," and it names a span of time that was bounded by mass extinction events. It was generally warm, with relatively little climatic differentiation between the equator and the poles, and it was the time of the dinosaurs.

The last 12,000 years—about the last 90 seconds on the Earth Calendar—are an epoch known as the Holocene. It started with the retreat of glaciers from North America and Europe (creating the Great Lakes between the US and Canada). The Holocene, like any epoch, consists of a range of changes in weather in different climatic zones around the planet. Earth is never static. Yet in the scheme of things, the Holocene has been a stable period that enabled humans to develop settled agricultural and industrial civilizations.

Humans have since become dominant forces on the planet. At first, this was a gradual process as we felled trees, drained wetlands, and used fire to shape ecosystems. Humans likely hunted some animals to extinction. Megafauna[2] extinctions, for example, follow a pattern that parallels the spread of humans into previously uninhabited areas, from Australia (c. 40,000 years ago) to North America (c. 13,000 years ago) to the Commander Islands[3] (250 years ago).

And then in just the past few hundred years, human impacts started accelerating rapidly. To give just one important indicator, it took 200,000 years for our population to reach one billion. It took just the last 200 years for it to skyrocket to eight billion. What is going on?

---

2   The large animals of a particular region at a particular time.

3   Russian islands, at the far west of the Aleutian Island chain.

The World Population over the Past 12,000 Years

Demographers expect rapid population growth to end by the end of the 21st century. The UN demographers expect a population of about 11 billion in 2100.

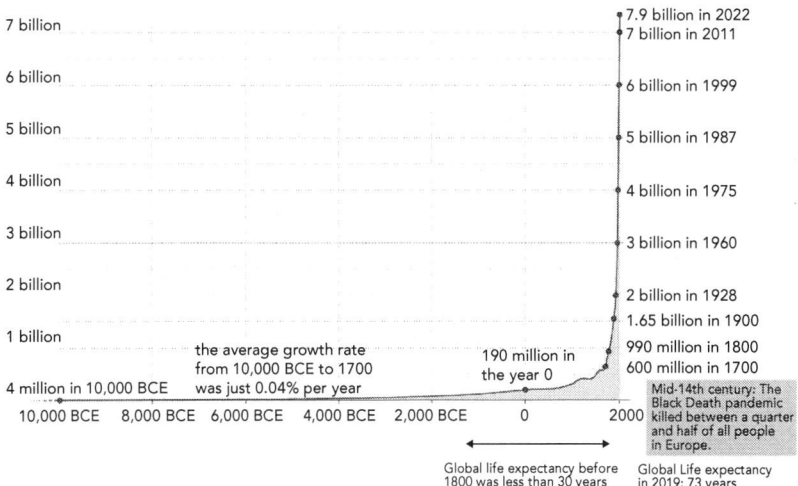

FIGURE 1.3

For a similarly remarkable trend, consider global Gross Domestic Product (**GDP**) over the past 2,000 years. The global economy didn't hit $1 trillion in value until 1800. It has since jumped rapidly to over $100 trillion. Technological and political-economic innovation in the modern age allowed societies to break out of the zero-sum logic of pre-growth economies known as the "Malthusian trap."[4]

In 2002, the Dutch atmospheric chemist Paul Crutzen penned a short article in the journal *Nature* titled "Geology of Mankind." Noting the escalating effects of humans on the global environment, Crutzen wrote, "It seems appropriate to assign the term 'Anthropocene' to the present, in many ways human-dominated, geological epoch." He thought that this new epoch could be dated to the human signature of growing concentrations of greenhouse gasses (**GHGs**) in the atmosphere. This began, he noted, at the time of James Watt's steam engine in 1784. Crutzen's idea has sparked ongoing debates about the Anthropocene or what others call The Great Acceleration.

---

4   In his 1798 book, Thomas Malthus argued that population growth must outstrip food production, thus leading to lower and lower living standards, and eventually starvation.

World GDP over the Past 2,000 Years

Total output of the world economy. This data is adjusted for inflation and differences in the cost of living between countries.

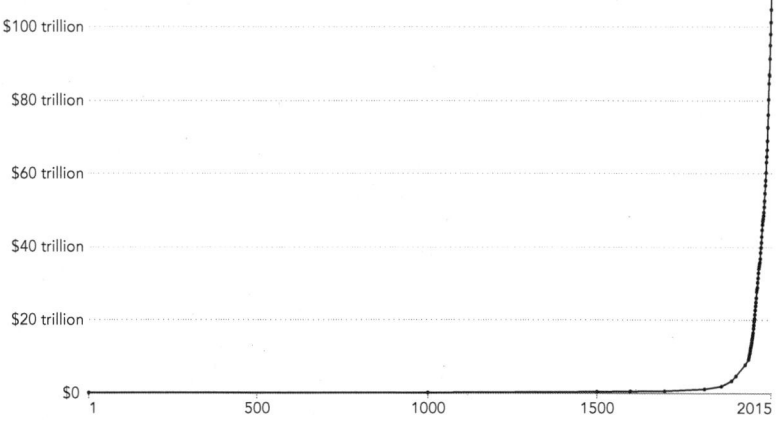

Note: This data is expressed in international-$[1] at 2011 prices.

1. International dollars: International dollars are a hypothetical currency that is used to make meaningful comparisons of monetary indicators of living standards. Figures expressed in international dollars are adjusted for inflation within countries over time, and for differences in the cost of living between countries. The goal of such adjustments is to provide a unit whose purchasing power is held fixed over time and across countries, such that one international dollar can buy the same quantity and quality of goods and services no matter where or when it is spent. Read more in our article: What are Purchasing Power Parity adjustments and why do we need them?

FIGURE 1.4

Signs of the Anthropocene

Imagine alien archaeologists visit Earth one million years from now (about two hours in the future on the Earth Calendar). Let's say that humans are gone (maybe we are living on Mars!). The aliens dig into accumulated layers of rocks until they reach our current time. Suddenly, they exclaim, "The age of humans!" What might they find? What are the telltale signs that would signify our dominant presence on the planet?

Not many things leave a sign that could last that long, making it hard to know just what the aliens would find. You might think, for example, that they would find a thin line of radiation deposited around the globe from the over 2,000 nuclear tests that have been conducted. Perhaps, but that radioactive material would likely have decayed by then. Nonetheless, consider a list of some of our impacts on Earth that might indicate that this is the "human age":[5]

---

5   Unless otherwise noted, the statistics in this section come from "Our World in Data," https://ourworldindata.org/. Readers might also consult Steffan et al. 2015.

1. *The technosphere.* This is the sum of all the materials that humans produce: concrete, aggregates (like sand and gravel), bricks, metals, asphalt, plastics, glass, etc.

   - The technosphere is growing rapidly. Globally, we are building the equivalent of a *new* New York City every month. Plastic hardly existed before 1950. Now, over 380 million tons are made each year.
   - A team of scientists estimated that the amount of human-made mass (1,154 gigatons) now exceeds the weight of all living biomass (1,120 gigatons) (Elhacham et al. 2020).
   - The technosphere does not recycle itself like the biosphere. Over two billion tons of solid waste are produced every year. That number is expected to keep rising.
   - Not all waste is well-managed. Eight million tons of plastics gets washed into the oceans annually. In addition to long-running concerns about pesticides and herbicides, there are new worries about "forever chemicals" (PFAS) that contaminate water with unknown health impacts. Indeed, rain water falling anywhere on Earth is contaminated with PFAS.

2. *Loss of biodiversity.* Earth has experienced five mass extinction events. Could the Anthropocene be another name for the sixth mass extinction?

   - This is hard to say, because we don't know much about most species on the planet. And only a fraction of species has been evaluated for extinction risk.
   - Nonetheless, there are worrying signs. Extinction rates today are much higher than background rates and higher even than previous mass extinction events. Around one million species are threatened with extinction.[6]
   - Habitat loss is a major factor in shrinking biodiversity. About 1,000 years ago, only 4 percent of habitable land was used for agriculture. Now 50 percent of all habitable land on Earth is devoted to growing food for humans.

---

6   IPBES. 2019. "Summary for policymakers of the global assessment report on biodiversity and ecosystem services of the Intergovernmental Science-Policy Platform on Biodiversity and Ecosystem Services." S. Díaz et al. eds. IPBES secretariat, Bonn, Germany. https://doi.org/10.5281/zenodo.3553579.

* If you weighed the biomass of all living mammals, 95 percent would be humans and our livestock such as cows. Only about 5 percent is wild mammals. Similarly, about 70 percent of the weight of all living birds would be chickens and other poultry. Over 50 billion chickens are slaughtered every year for human consumption. Maybe the aliens will find piles of chicken bones all over the planet! Would they call it "the Chickenocene"?

3. *Altered landscapes, oceans, and nutrient cycles.* Deforestation, farming, mining, and dam-building activities change sedimentary processes in ways that have enduring geological impacts.

   * In the past 300 years, people have cut down 1.5 billion hectares of forests, nearly one third of all forested land. Rates of deforestation peaked in the 1980s, but roughly 10 million hectares of forests are still lost annually, mostly in the global south. Globally, wetlands are disappearing even faster than forests.
   * Emissions of CO_2 and other GHGs[7] are warming the planet, leading to sea-levels that are higher than at any point during the Holocene. They also have the effect of acidifying the ocean water, which can lead to the dissolution of calcium carbonate in coral reefs.[8]
   * The Green Revolution in agriculture starting in the 1950s increased global crop yields. Fertilizer use has since altered the chemistry of the planet. Levels of nitrogen and phosphorous in soils have doubled in the last century. These nutrients are present in the environment at levels far exceeding any time during the Holocene.
   * Human mining, construction, and extraction activities move more material than the sediment carried by all the rivers of the world.

4. *Changed climate system.* GHG emissions are changing the climate in ways that will have lasting impacts.

   * Burning fossil fuels is dumping enormous amounts of heat energy into the climate system, much of it absorbed by the oceans. It is the

---

7   Greenhouse gasses—that is, gasses that trap heat in the atmosphere.

8   The structures produced by colonies of tiny coral animals are made of calcium carbonate (chalk). This dissolves in an acid solution.

heat equivalent of seven Hiroshima atomic bombs exploding in the oceans every second (Abraham 2022). *Every second.*

- The global average temperature is roughly 1.2°C higher than when record-keeping began in the nineteenth century. Atmospheric concentrations of CO_2 are over 420 ppm, far exceeding Holocene levels. Methane (CH_4) is a more potent GHG than CO_2, and accounts for roughly 20–30 percent of climate warming since the Industrial Revolution. Atmospheric concentrations of CH_4 have doubled in the past 200 years and also far exceed Holocene levels.

- As we will learn, increased temperatures are having several impacts, including melting glaciers, rising sea-levels, and more frequent and intense storms. All of these impacts can leave their marks on the geological record.

What to make of all this? The alarmed among us might say that this is a recipe for disaster. There are so many big, rapid changes pushing natural systems beyond the limits of the Holocene, which is the range of conditions to which human civilization is adapted. To the above list, we can add nuclear weapons proliferation, a host of disruptive technologies (including AI [Artificial Intelligence]), and increased risks of novel diseases and pandemics. It is a multi-dimensional situation of precarity, radical uncertainty, or what some call a "polycrisis."

But the cautious and concerned among us might note that civilization keeps on progressing. In many parts of the world, pollution is decreasing as wealth and life-expectancy increase. Fewer people are dying from weather- and climate-related disasters. Renewable energy and other climate-helpful technologies are accelerating faster than most experts thought possible. Maybe alarmism is overblown, because we are getting smarter and wealthier, enabling us to solve problems before they become catastrophic.

The big numbers describing the Anthropocene don't speak for themselves. What story do you make of them? What is your mental model?

Who Is This "We"?

The term "Anthropocene"' naturally invites talk of "humans," which then turns into a way of talking about "we"' and "our" and "us." It tells a story of *Homo sapiens* as a species relatively weak in muscle power but endowed with big brains. Over time, we figured out ways to control nature in order to improve human health, safety, and comfort. There's that "we" again.

Yet, just as biodiversity has been eroding, so has cultural diversity. Consider language as an indicator of culture. Languages have always gone extinct, but like living species, languages are also going extinct at an accelerating rate recently. Between 50 percent and 90 percent of remaining languages face extinction by 2100 (Austin 2011). In North America, the decline of native languages has been driven in large part by war, genocide, and policies of forced cultural assimilation implemented through boarding schools and other means.

The Anthropocene, then, is not the spread of all forms of humanity. It is not about all peoples coming to dominate Earth. Rather, it is a story about the spread of a certain way of being human. It is a way that pushes other ways of life to the margins. This is a story about biocultural homogenization (Rozzi 2018), the loss of biological and cultural diversity.

So, not all humans are equally responsible for the so-called Anthropocene. Let's take CO_2 emissions as an example. Consider the same chart from the introduction but now disaggregated by region.

Annual CO₂ Emissions by World Region

This measures fossil fuel and industry emissions.[1] Land use change is not included.

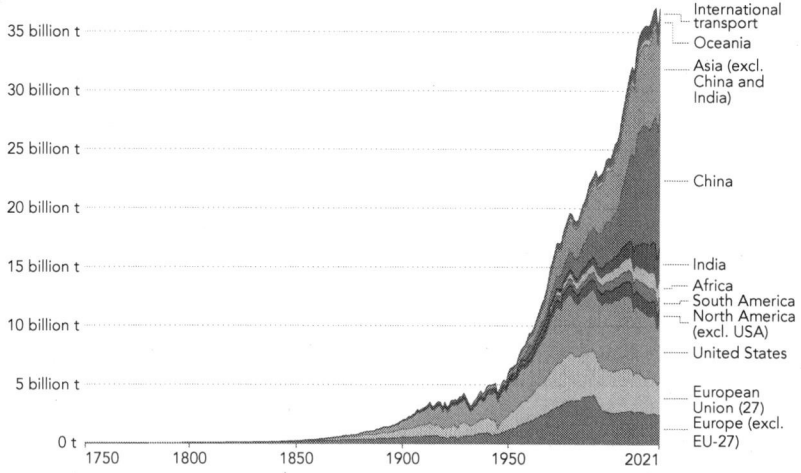

1. **Fossil emissions:** Fossil emissions measure the quantity of carbon dioxide (CO_2) emitted from the burning of fossil fuels, and directly from industrial processes such as cement and steel production. Fossil CO_2 includes emissions from coal, oil, gas, flaring, cement, steel, and other industrial processes. Fossil emissions do not include land use change, deforestation, soils, or vegetation.

FIGURE 1.5

Clearly, emissions are extremely unequal. Africa is home to 16 percent of global population but accounts for just 4 percent of emissions. North America, by contrast, is home to 5 percent of total population but accounts for about 18 percent of emissions. The per capita emissions of the average North American is 17 times higher than the average African and 5 times higher than the global average.

Of course, there are also differences *within* nations. Wealthy Americans, for example, produce more emissions than those in the middle and lower classes. And we can parse emissions in totally different ways. For instance, a 2017 Carbon Majors report estimated that just 100 companies were linked to 71 percent of global industrial GHG emissions since 1988 (Griffin 2017). Not surprisingly, this list includes lots of state- or investor-owned oil, coal, and gas companies like Saudi Aramco, Gazprom, National Iranian Oil, ExxonMobil, Shell, and BP. Questions of identity and responsibility are tricky. For example, should we think in terms of nation states, companies, individuals, or some other category? And who, really, is a company? Can they be separated from the people who consume their products?

More broadly, if the Anthropocene is not the work of all humans equally, then is there a better way to diagnose what or who is causing these massive impacts? Who is this "we" hidden behind the supposedly universal *anthropos*? Consider two clusters of theories.

1. *The West, Modernity, and Colonialism.* Industrial CO_2 emissions first started climbing in the UK and other parts of Europe in the eighteenth century. As this was the beginning of the big impacts we listed above, perhaps a better name for our geological epoch is the "Eurocene" or the "Modernocene"—to indicate the centrality of European nations or modern industrial-scientific-political structures. These terms would focus our attention not on humanity but modern European nations and corporations and the historical period in which they came in contact with other cultures and shaped the entire biosphere by exporting and imposing their way of life. Because Europe and North America are often called "the West," some call this process "the Westernization of the world" (Latouche 1996). Others point to settler colonialism, extractive colonialism, or the "colonial matrix of power" as the essential engine of the so-called Anthropocene (Mignolo 2011).

2. *Capitalism and Neoliberalism.* In a critique of Crutzen's "Geology of Mankind" article, Andreas Malm and Alf Hornborg (2014) similarly argue that it is not "humanity" that has been acting like a swarm of locusts consuming resources and putting an end to the Holocene. Rather, it is a powerful system and set of ideals that orient the human condition around *growing production and consumption.* This system is capitalism. So, this point in geologic time should really be called the Capitalocene to indicate the dominance of this way of organizing the world. Others agree, but emphasize post-World War II growth and the creation of neoliberalism, a version of capitalism that emphasizes privatization, deregulation, austerity, and high-levels of management usually disguised as "free markets" (Klein 2007).

These theories attempt to diagnose our unprecedented epoch in ways that are more accurate. In everyday discourse, we tend to just talk about *the economy* and we call its growth and spread *development.* And we tend to assume it is all in the name of human *well-being.* In the past few hundred years, the economy became more strongly disembedded from local and regional areas. It started globalizing in chains of supply and demand. What the theories above emphasize is that this process is marked by deep inequalities between colonizer and colonized and between capitalist and proletariat (those who own the means of production and those who only own their labor power). To pick a salient example, there are children working in the cobalt mines in the Democratic Republic of Congo and there are Americans driving $100,000 Tesla cars that run on the batteries made with that cobalt. Of course, the sites of production are often hidden from consumption, making it easy for people to not think about the impacts of their actions.

The growing global economy is a morally ambiguous story. It is stained by violence, slavery, genocide, imperialism, cultural erasure, and forced dispossession (removing people from their lands). It creates new kinds of poverty tied to forced minimum levels of consumption and, as noted above, it has taken a heavy toll on the environment. GHG emissions, for example, have historically been tightly coupled to Gross Domestic Product (GDP), the main measure of economic development. Wealth and emissions have gone hand in hand. We'll see how decoupling them is key to climate policy. Development tends to create new needs, not just satisfy a given set of needs. After a certain point of development, people tend to not be any happier even though they have more and more stuff.

Yet this is also a story about rising standards of living, improved economic opportunities, rising access to education, expanded political and civil liberties, and reduced hunger. The United Nations has codified the beneficial aspects of development in terms of 17 Sustainable Development Goals (SDGs), which are each further specified in terms of targets and indicators. For example, the goal of "end poverty in all its forms everywhere" has one target of eradicating extreme poverty by 2030, which is tracked by using the indicator of the proportion of population living below the international poverty line. These SDGs frame the larger context of climate politics.

One more fundamental point about economics should be stated right away: *growth is king.* Underneath partisan political polarization, this is something nearly everyone agrees on. Politicians will debate *how* to stimulate economic growth (e.g., tax policy) or what kind of growth to favor but they rarely debate *whether* to promote economic growth. Economic growth is about jobs, innovation, financial security, and opportunity. Stock markets and GDP figures stand in as metrics for progress and well-being. Periods of economic contraction, like during the Covid-19 pandemic, show the kinds of hardships that follow from stalled growth—hardships that often disproportionately impact the poor and marginalized.

This is to say that the economy is foundational to all of politics, including climate politics. We ought to beware the temptation to think of the economy as somehow a value neutral sphere, as if the "private" sector did not have very public ramifications or as if the "free market" wasn't actually the product of decisions we have made about how to distribute power.

So much of our political activity is a structure built atop the economic bedrock below. Indeed, politicians are often at the mercy of economic imperatives. This is true at the local level where, for example, city councils usually only have little control over the forces of development like new residential construction. And it is true at every level all the way up through national government policies, over which huge corporations sometimes have inordinate power. Yet it is also true that "private" forces of the economy are guided and constrained by public laws—governments and the people they represent can and do shape the trajectory of development.

Why Are We?

Call it the Great Acceleration, the Anthropocene, or the Capitalocene. By any name, it is an age of massive change and rapid growth with indicators on graphs curving sharply up. What values are driving all those curves upward? In other words, what is all this consumption and production *for*? Is there a guiding vision of the good life or well-being?

We could say that this is just what happens when humans tap into power dense materials like fossil fuels. All that energy is bound to push things beyond old boundaries. There must be some truth to this, but it seems inadequate for similar reasons discussed above. Other cultures had long used coal and petroleum without the same accelerating global growth. Even in parts of pre-modern Europe, coal was often utilized without the massive changes that we have been examining.

The modern world that has sparked the Great Acceleration required new ideas and values, not just the right material conditions and machines. There were new ideals of human flourishing, a new conception of our relationship to God and the natural order, an emphasis on the individual above the communal, and a new way of picturing the world. In modern Europe, nature came to be seen as a set of resources to be manipulated by human labor in order to unlock more value and steer it toward more useful ends. But toward what ends? Development toward *what?*

The philosopher Albert Borgmann (1984) offers one way to think about this. He thinks of development in terms of *convenience* and "the device paradigm," which is a predominant, but not universal, pattern of life. Most of us are living within the device paradigm. There are, though, folks who live "on the land" or "off the grid" in ways that break with the ruling pattern. Economic development is the process of replacing these other ways of life with the device paradigm. You might think of the small ranch giving way to the industrial-scale confined animal feeding operation (CAFO) or the local shoemaker being displaced by Walmart.

Let's think first about those less technologically intense and more place-bound ways of life characteristic of the "undeveloped" world. Borgmann uses the example of a wood-burning stove by way of contrast with the device paradigm. With such a stove, he imagines someone chopping wood from a nearby lot, stacking it, hauling it in on cold mornings, and lighting the fire. The experience is embedded in the context and it takes skillful engagement. That is, the materials needed for the heat are locally-sourced, and one needs some know-how to make things work. The stove serves as a center that gathers the family around its warmth. Thus, a way of life filled primarily with "focal things" (like stoves) and "focal practices" (like tending to it) is one where means and ends are experienced together within the context of a specific place. It is generally slower-paced with fewer possessions.

By contrast, the modern central heating system is a typical example of a "device." It is serviced by wires and pipelines bringing electricity or gas from networks of extraction, processing, and transport. So, there is a much stronger separation between means (the hidden background of machinery) and ends (the foreground experience of warmth). There is little demanded by way of skillful engagement, strength, or attention, because the temperature is adjusted at the push of a button. Such a device is disembedded from any local context. Recall as well the example of the Tesla driver as the foreground of consumption (ends) and the child in the cobalt mine as the background of production (means).

Now consider other examples to see how devices more generally pattern the predominant global culture. Think about packages of cellophane-wrapped meat at the grocery store or on-demand streaming entertainment or consumer goods purchased via Amazon and shipped to your door. Consider especially how the means might radically shift without altering the experience of the ends. For example, solar panels can bring you the same light as a coal-fired power plant. Notice how very little most of us understand about the vast background of production that makes our device-filled lives possible. And consider how little we tend to think about, let alone experience, the far-flung impacts our device-filled lives have.

So, the device paradigm is a way of life that:

a. Separates means (production/labor) and ends (consumption). The means are largely hidden from daily experience while the ends, the commodities, are foregrounded.

b. Explains the way we experience the de-contextualizing or disembedding forces of development. We find the same pattern in different places. Place matters less and less, because commodities can be supplied anywhere from the network of production. Thus, the phenomenon described above as biocultural *homogenization*.

Admittedly, we often have an abundance of choices living in a world patterned in this way. So, in some sense, this way of life creates diversity, mostly a diversity of consumer brands. Note, though, that these are choices *from within* the device paradigm. We are not offered the choice of whether to adopt the device paradigm itself. It is presented as the basis for choice, not itself something we can choose (unless you are prepared for radical life changes). You can choose your car, for example, but not whether to live with cars and highways. And this is because the device paradigm, as the outcome and goal of development, is widely assumed to be good. It supplies our predominant, unspoken ideal of well-being.

Now we can see one answer to the question: what is all this growing development for? Borgmann calls it "the promise of technology" to liberate and enrich our lives. *It is for security, comfort, and convenience.* We can be disburdened from the hardships of season, place, disease, hunger, and toil. It is hard, after all, to tend to the stove. And we can have reliable access to an abundance of goods and services. This was the key idea at the birth of the so-called Anthropocene: to control nature in order to make human life more secure and convenient. Indeed, the pattern of development resulting from this idea literally *convenes* material goods and services for us. In Borgmann's terms, it makes commodities "available" in ways that are instantaneous, ubiquitous, safe, and easy.

The development model tends to secure vital commodities (food, water, and energy) with great reliability. This is clearly a good thing. Yet, this reliability invites thoughtlessness, because the systems that sustain us are out of sight and out of mind. For example, you just throw garbage "away." You don't have to know where your groceries come from. The water just comes out of the tap in the kitchen sink. Convenience, then, tends to disburden us not just from physical labors but also from thinking and, arguably, from our moral responsibilities. And although the device paradigm brings freedom in one sense, it also brings a kind of dependence on machines, systems, and institutions. Maybe a more thoughtful life at a slower pace with fewer things is a more viable path toward well-being. In Chapter 4, we'll return to this growing pattern of security and convenience to ask whether it can and should be sustained.

Conclusion

Humans are newcomers to this "pale blue dot." We have built civilizations that are adapted to climate conditions that we are now altering. Our very success in the development project is posing the dangers of climate change. But this "we" and "our" is fraught moral language. Not everyone has created or benefited from development equally. Not everyone is equally exposed to the dangers of a changing climate. So, as we aim for "sustainable development," who has the responsibility to do what? We think of development as progress—as a brighter future—so it is important to critically consider these terms. What is a good life and a just society? What is all this growing development for?

Activities and Questions

1. Watch the movie *Wall-E*. How does it portray the Anthropocene? How does it speak critically about development: both to the recklessness and wastefulness of its means as well as the dubious value of its ends, that is, the ideal of well-being that guides development?

2. What other critiques of the Anthropocene can you think of? Research the post-humanist, feminist philosopher Donna Haraway. What does she mean by the "Chthulucene"?

3. If progress is not measured by a growing GDP, what metrics can we use? Research the Human Development Index and the Gross National Happiness Index. We think of happiness or well-being as the ultimate good, but what is it and how do we measure it?

References

Abraham, John. 2022. "We Study Ocean Temperatures. The Earth Just Broke a Heat Increase Record." *The Guardian*, January 11.

Borgmann, Albert. 1984. *Technology and the Character of Contemporary Life.* Chicago: University of Chicago Press.

Crutzen, Paul. 2002. "Geology of Mankind." *Nature* 415 (January): 23.

Elhacham, Emily, et al. 2020. "Global Human-Made Mass Exceeds All Living Biomass." *Nature* 588 (December): 442–44.

Frodeman, Robert, and Mark Bullock. 2022. "We're Children of Ice and Snow. Can We Survive the Coming Heat?" *Psyche*, April 5. https://psyche.co/ideas/were-children-of-ice-and-snow-can-we-survive-the-coming-heat.

Griffin, Paul. 2017. *The Carbon Majors Database*. London: CDP Worldwide. https://cdn.cdp.net/cdp-production/cms/reports/documents/000/002/327/original/Carbon-Majors-Report-2017.pdf.

Klein, Naomi. 2007. *The Shock Doctrine: The Rise of Disaster Capitalism*. New York: Picador.

Latouche, Serge. 1996. *The Westernization of the World*. Cambridge, MA: Polity Press.

Malm, Andreas, and Alf Hornborg. 2014. "The Geology of Mankind? A Critique of the Anthropocene Narrative." *The Anthropocene Review* 1 (1): 62–69.

Mignolo, Walter. 2011. *The Darker Side of Western Modernity*. Durham, NC: Duke University Press.

Rozzi, Ricardo. 2018. "Biocultural Homogenization: A Wicked Problem in the Anthropocene." In *From Biocultural Homogenization to Biocultural Conservation*, vol. 3, *Ecology and Ethics*, ed. Ricardo Rozzi et al. Cham: Springer.

Rozzi, Ricardo, and Alexandria Poole. 2020. "The '3H's' of (Habitats, Habits, Co-in-Habitants) of the Biocultural Ethic: A 'Philosophical Lens' to Address Global Changes in the Anthropocene." In *Global Changes: Ethics, Politics and Environment in the Contemporary Technological World*, ed. L. Valera and J.C. Castilla, 153–70. Cham: Springer.

Sagan, Carl. 1994. *Pale Blue Dot: A Vision of the Human Future in Space*. New York: Random House.

Steffan, Will, et al. 2015. "The Trajectory of the Anthropocene: The Great Acceleration." *The Anthropocene Review* 2 (1): 81–98.

CHAPTER 2

What Is Climate Change?

Yearly Global Surface Temperature and Atmospheric Carbon Dioxide (1850-2022)

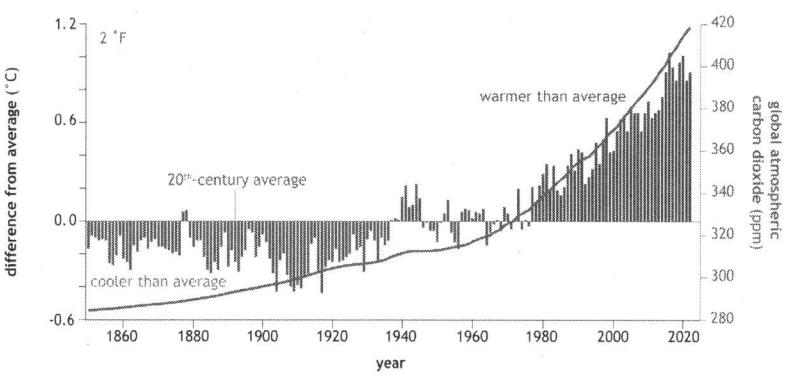

FIGURE 2.1

The writer Mark Twain once quipped, "Climate is what we expect, weather is what we get." We "get" weather, meaning that is what we actually experience—shivering in a blizzard, sweating in a heatwave, or sheltering from a thunderstorm. And we "expect" climate, meaning that it establishes our way of making sense of the otherwise unconnected barrage of weather events. The climate is a framework for setting expectations about the future. We expect the climate every time we plant seeds in the spring or when planners allocate the usage rights to water in a river for years to come. Weather is the immediate (un-mediated) stuff of experience. Climate is always mediated—through statistics, scientific instruments, stories, technologies, and culture. Indeed, climate is mediated through the weather.

The climate says something about our need for durability, normalcy, patterns, and predictability. "The climate" is a structural component of our mental models discussed in the introduction. The philosopher Mike Hulme

(2017) argues that, "Climate is weather which has been cultured, interpreted and acted on by the imagination, through story-telling and using material technologies" (11). All cultures have invented ways to conceive of "the climate" and they have developed what Hulme calls "weather-ways," patterns of behavior, dress, emotions, practices, and memories that define a way of life in a particular climate. When I lived in Minnesota, winters were about heavy coats, snowplows, and ice fishing. I live in Texas now, where most folks don't own a winter coat and we have no snow plows.

Many of these "weather-ways" are changing. The ice fishing season in Minnesota, for example, has decreased by about 14 days over the past 50 years. The number of 100-degree days in Texas is expected to double in the next decade. The predominant pattern of development we examined in the last chapter complicates the story about weather-ways. As we saw, development tends to extract culture from place. The same highways, internet, electricity lines, and supply chains crisscross Texas and Minnesota. You can find many of the same foods, kinds of buildings, and behaviors. Being less dependent on a local climate zone means that we tend to be less aware of how it is changing. This begins to indicate the importance of Indigenous knowledge, because indigeneity (meaning "in-born" or "sprung from the land") is like the converse of development.

Hulme notes that all cultures have told stories about how and why climates change, stories that often feature responsibility, guilt, and supernatural forces. The predominant global culture of development that we examined in the previous chapter has an accompanying global scientific community that tells our story about a changing climate. This story is naturalistic, that is, it appeals to natural forces like solar radiation rather than supernatural deities. Yet this story is also humanistic, because it speaks to impacts on our lives and to the moral and political decisions that we make or fail to make. Our story too is about guilt and responsibility.

So, what is the climate? We can start to see that this is a physical and metaphysical question. The climate is about a material reality as described and explained by the sciences. And the climate, metaphorically, is about our imaginative realities as expressed in our cultures. This is why we also talk about a certain "political climate" or a new "intellectual climate." The idea of climate is connected to mood and *zeitgeist* or the spirit of the time and place.

To ask "what is" questions is to ask about reality. The climate exists, but what is its mode of being, how does it appear, or in what ways is it real? And how can we acquire knowledge about it? In this chapter, we'll begin with the modern scientific account of climate and climate change. Then, we will turn to everyday or commonsense experience to explore the human significance of climate change by asking what it is *like*. This will lead us to the theory of hyperobjects, which is a powerful way to connect the scientific and human accounts of the climate.

The Scientific Account

Let's first look at the scientific account of "climate" before turning to "climate change."

What Is Climate?

Etymologically, the word "climate" comes from the root *klei*, meaning "to lean." The Earth leans on its axis at a 23.5-degree tilt, meaning that some areas receive more energy from the sun and some receive less. Ancient natural philosophers (before the term "scientist") divided the Earth into *clima* or zones based on the angle of the sun. They thought there were perhaps 20 or 30 such zones, changing with the quality or angle of sunlight as one travels north. By around 1600, temperature and precipitation had become more important to these early geographers, such that they began examining the weather associated with different regions.

Modern science has refined these early accounts into various climate classification schemes. The Köppen–Geiger system, for example, features five main climate groups with further sub-divisions from there. Such zones often correlate to biomes or ecosystems, like the sugar maple forests in central Minnesota or the mesquite savannahs of north central Texas. Climate zones need not be large. Despite its small size, for example, Hawaii features a wide diversity of climate zones.

Here are a couple of contemporary definitions of climate, one from climatologists and the other from Earth scientists.

1. *Climate as long-term average weather.* Figure 2.1 is a good example of this understanding of climate, because it focuses on how temperatures differ from the average. We often talk about climate in terms of average (or typical) weather. For example, Phoenix has a hot, dry climate, while Amsterdam is cool and wet. Climate change means a shift in the

long-term averages: places get hotter or wetter, for example. Climate understood in this way can be studied as a statistical phenomenon about variables such as temperature, precipitation, and wind. Scientists usually take a span of thirty years of weather data to denote the climate. It's like baseball: the weather is like each time a baseball player goes up to bat and the climate is like the long-term average of all those batting attempts. But be cautious with this analogy. Climate is *like* a batting average and can be studied using statistics, but it is a complex, dynamic, and chaotic system that is not as simple as baseball and certainly not as simple as the statistical phenomenon, say, of a game of dice.

2. *Climate as the interaction of Earth's systems.* Here we think more about energy than statistical averages. The interactions of the biosphere (life), lithosphere (rocks), cryosphere (ice), atmosphere (air), and hydrosphere (water) give rise to the climate system. It is an enormous flux of energy and materials, powered by radiant energy from the sun and the heat at Earth's core. Climate change denotes significant changes in those interactions, like when temperatures fall, glaciers advance, and sea-levels drop. Here we can also capture the sense of climate change as new extremes and not just new averages. More heat in the system can power more extreme weather events. And this can mean big changes sometimes happen suddenly, like a 2016 early ice breakup that caused 10,000 emperor penguin chicks to drown, because they were not yet ready to swim. Sudden events like this can decimate populations and even entire species.

The Earth's atmosphere is key to the climate system, especially in regulating temperature. Lacking an atmosphere, the moon experiences wild temperature swings over the course of a lunar day from around +200 to -200°C. Solar energy entering the Earth's atmosphere is eventually radiated back into space. Some of the outgoing energy, though, is reabsorbed by greenhouse gasses (GHGs) and radiated back down to the surface of the Earth in the "greenhouse effect," which warms the planet. As we emit more GHGs into the atmosphere, it increases the Earth Energy Imbalance (EEI), which is the difference between the amount of energy from the sun arriving at the Earth and the amount returning to space. An increasing EEI is a scientific way to describe global warming, and scientists can measure this with a fair degree of precision. The EEI roughly doubled between 2012 and

The Climate System

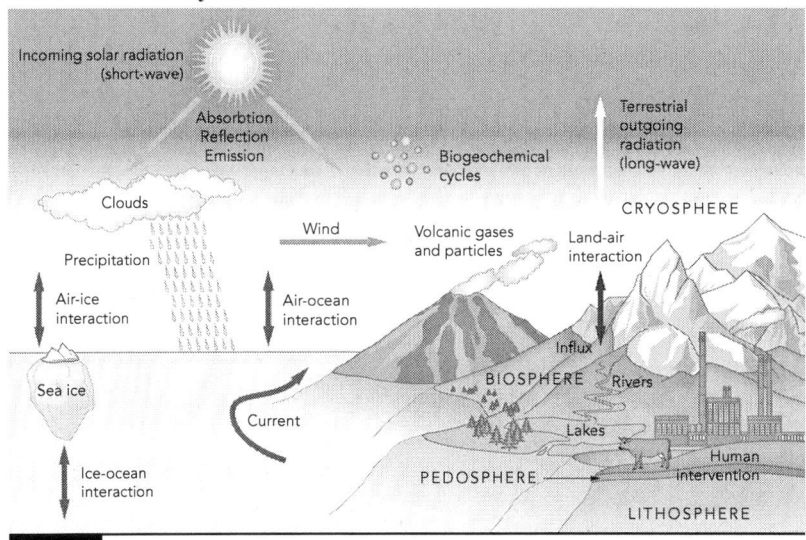

FIGURE 2.2

2023, which represents a staggering amount of additional heat energy on the planet (see Hansen, Sato, and Ruedy 2023)

There is a tight correlation between atmospheric GHG concentrations and temperature (as Figure 2.1 illustrates). Of particular concern are carbon dioxide (CO_2) and methane (CH_4) because they are emitted in such enormous quantities, but other GHGs like hydrofluorocarbons (HFCs) are also important. In geological history, even relatively slight variations in CO_2 levels have meant the difference between glaciers over New York and steamy rainforests in the Arctic.

Other parts of the climate system interact with the atmosphere in carbon cycles. The biosphere, oceans, and lithosphere act as carbon sinks when they absorb more carbon from the atmosphere than they release. These sinks or reservoirs help to reduce the warming effects of human emissions. The IPCC (2021) projects with "high confidence" that land and ocean sinks will become less effective at storing anthropogenic CO_2 emissions over time, meaning that a higher proportion of our emissions will remain in the atmosphere.

What Is Climate Change?

Think of the climate as the unchanging parameters within which the changing weather occurs. Take Amsterdam, for example. It gets cold and wet in the winter and then warmer and drier in the summer. The tides of the North Sea rise and fall. So, things change, but they change in pretty much the same way. One useful concept here is *stationarity*, which means that the properties that give rise to change are themselves unchanging. Stationarity means that there is no trend, that is, there is *constant variance* over time: there are changes but the changes are consistent and do not themselves change. Or you can think of stationarity as change that happens within unchanging, normal, predictable parameters. The temperature goes up, but not past a certain point, and down, but not past a certain point. The tides come up, but only so high before going down again but only so low.

 ▪ Climate change, then, is when the parameters themselves change. *Climate change is change in the way things change.*

Low temperatures, for example, consistently don't get as low and high temperatures consistently get higher. The high tide gets higher than it used to. There is a new normal, a new kind of "lean" that puts things out of alignment. This is how, over time, a place that was once cool might become tropical. It is how places that were once dry land get swallowed by the sea.

Climates always change and have changed naturally for billions of years. My home in Texas used to be under a shallow sea. Before that, this part of Texas was not even in the same place on the globe, because it is on a tectonic plate moving around. In many cases, adapting to a changing climate makes sense, regardless of whether humans are driving the changes or not.

Anthropogenic (human-caused) climate change is our main concern, because "we" (who?!) are responsible due to our GHG emissions. It is also concerning, because the changes are happening very rapidly. Of course, they seem slow compared to, say, video games or stock market fluctuations—that's a big part of the problem: climate change is not experienced with the same urgency as a tornado or hurricane. The pace of anthropogenic climate change is extremely fast when considered from the viewpoint of the Earth calendar. More troublingly, it often outpaces our ability to react with new policies and new infrastructure.

We are digging up the lithosphere and combusting it into the atmosphere. And this is happening faster than terrestrial and ocean sinks can absorb the emitted GHGs. As a result, things are changing in new ways.

Consider five examples.

- In June 2021, temperatures topped 48°C (120°F) in British Columbia, a place where almost no one has air conditioning. This was so far beyond the observed experience that it exceeded even the statistical models' outermost potential extremes for that area. That is the end of stationarity, or the end of our usual *expectations* about climate. As one meteorologist put it about a heatwave in China in 2022, "there is nothing in world climatic history which is even minimally comparable" (Wallace-Wells 2022).

- The Australian Bureau of Meteorology uses a series of colors to represent temperatures on weather maps. For a century, their map topped out at 50°C (122°F). Yet a heat wave in 2013 compelled the bureau to add a new color for temperatures up to 54°C (129°F). That new color on the weather maps represents a change in the way things are changing.

- In March 2022 a heatwave in Antarctica set a new world record for the largest temperature excess above normal ever measured at an established weather station (+38.5°C/+69.3°F). At the same time, a heat wave in Pakistan caused a "glacial lake outburst flood."

- The American Southwest has been in a severe drought for so long that some scientists think the term "aridification" is more appropriate. A "drought" sounds temporary, like weather. You expect things to go back to normal after a drought. But "aridification" means a long-term climatological shift. Despite a very wet 2023, water levels in the Colorado River are unlikely to ever return to what we expected as normal for the past century. Scientists have attributed this aridification partly to anthropogenic causes (see Overpeck and Udall 2020).

- As an El Niño[1] took hold, the summer of 2023 witnessed several historically unprecedented extremes, including wildfires in Canada, heat waves in Mexico and Pakistan, record high sea surface temperatures in the North Atlantic, and record low sea ice extent in Antarctica. September 2023 was the hottest September ever recorded.

---

1   An area of warm water that appears in the Pacific every few years, affecting the climates of many areas, especially on the Pacific coast of the Americas.

It was 0.46°C (0.83°F) hotter than the previous hottest September. This smashed the record for monthly global temperature anomaly, which had been 0.09°C (0.16°F).

Of course, black swans or rare events happen. But many such extreme events are happening now with increased frequency and intensity because we have super-charged the atmosphere and oceans with more heat. One black swan is an outlier, but a whole flock is the new normal. If extreme events like the ones described above keep happening, then over time the mean will change, which is a shift in the climate. One way to represent this schematically is with overlapping bell curves[2]—climate change here is a shift from one probability distribution of weather events (temperature or precipitation, for example) to another probability distribution.

Climate Change Bell Curves

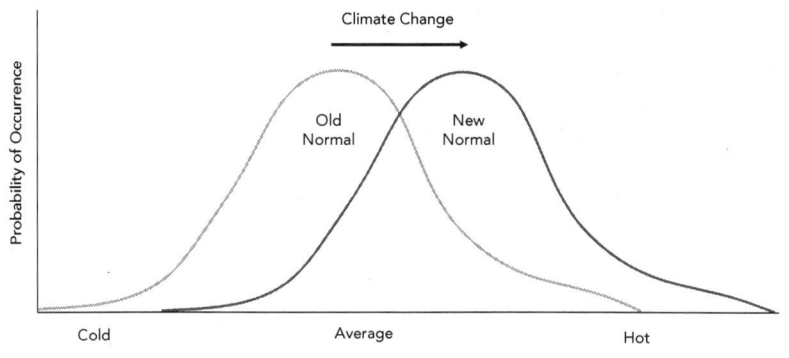

FIGURE 2.3

We are changing the way things change. By altering atmospheric GHG concentrations, we are creating a new normal, a new climate. What we once called "extreme" events are now becoming more commonplace, though we need to always be careful to parse historical data when making such claims in specific cases (UN 2022).

The best place to turn for a comprehensive, global, and systematic accounting of climate change risks, damages, and impacts is the IPCC

---

2   A bell-shaped curve on a graph showing a typical probability distribution: low on extreme ends, high in the middle.

Working Group II reports. Here, the IPCC tracks over a hundred climate-related hazards. Three things are critical to note about the risks and impacts of climate change.

* First, adverse impacts, losses, and damages escalate with every increment of global warming. So, near-term mitigation of GHG emissions is vital.

* Second, adverse impacts can also be reduced through efforts that decrease vulnerabilities. As we'll see, this is why adaptation and finance are vital, especially for poor regions of the world where vulnerability to climate risks is high.

* Third, climate risks are complex and increasingly difficult to manage. Many hazards occur simultaneously and they interact with non-climatic risks in ways that create compounding risks that cascade across multiple sectors of society—finance, health, energy, etc. Keep this in mind when we learn about hyperobjects below!

In a 2021 report, the IPCC Working Group I noted, "Human-induced climate change is already affecting many weather and climate extremes in every region across the globe" (10). And since their report seven years earlier, "Evidence of observed changes in extremes such as heatwaves, heavy precipitation, droughts, and tropical cyclones, and in particular, their attribution to human influence, has strengthened" (10). As a result, *the world we have built is not well-adapted to the planet we are shaping*. Things are getting out of alignment. As the IPCC puts it, "Climate change represents an urgent and potentially irreversible threat to human societies and the planet" (Allen et al. 2018, 79).

The rest of the book will cover lots more about climate science and climate change. One other thing is important to address early on, though. My students often ask me why climate change is such a big deal when the shifting average temperatures we are discussing seem so small. After all, as we'll see in the next chapter, the goal is to keep temperatures from rising 1.5°C or 2°C above pre-industrial averages. Those seem like tiny numbers. I mean, on any given day the temperature might swing 10°C or more.

To appreciate why seemingly small numbers are such a big deal, you have to consider the vast size of the Earth's climate system, which stretches from the depths of the oceans to the upper reaches of the atmosphere. Built into the climate measurements like average temperature, then, are billions of calculations taken from around the world over long stretches of time. Shifts to the long-term average of such a huge system can only come about through the accumulation of more and more individual extreme measurements. Hidden behind that seemingly small change, then, will be massive, prolonged heatwaves and other extreme events. And these are what impact humans and the things that we care about.

There is another way that climate scientists sometimes convey this point, and it is a perfect transition to our next section, because it is an analogy. We might say that global warming is *like* a fever for the planet. A roughly 1°C (or 2°F) increase in our body temperature is noticeable: we start to feel lousy. Similarly, such a seemingly small shift for the climate is also troubling. Each increase of 1°C, for example, is likely to mean a 5–15 percent reduction in the yields of major crops and a 3–10 percent increase in the amount of rain falling in the heaviest events (Hayhoe 2019).

The Human Experience of Climate Change

We don't just study the climate as a phenomenon of scientific curiosity. Our sole objective is not contemplation or knowledge for the sake of knowledge. Rather, we want to know *what to do*. This introduces a puzzle, though, because science alone does not guide action. We also require values—things we care about that motivate us to act.

Above, I quoted the IPCC as writing: "an urgent and potentially irreversible threat." Is that what climate change *is*? The claim is an interpretation and a conclusion drawn from climate science, but it is not itself a scientific statement. It is a way to make meaning out of the numbers. We have to do such meaning-making activities—it is just part of the human condition. This means that art, stories, and political reasoning are just as essential to climate literacy as science is. Humans don't process information through a string of if-then or binary operations like a computer does. Rather, we make meaning through language and interpret the world through various cultural lenses and our mental models.

For this reason, when we think about what something *is*, we often have to think by analogy about what it is most *like*. This is how we often form connections in ways that help us to understand a phenomenon. This kind of *analogical reasoning* is key to art, poetry, ethics, politics, and law. It's central to how we form our mental models. We'll talk more about this in later chapters. For now, let's consider the 2021 Netflix movie *Don't Look Up* as a way to both practice and appreciate analogical thinking when it comes to asking what climate change *is*.

In the film, a pair of scientists discover a comet that will crash into the Earth in precisely 6 months and 14 days. It will destroy civilization. Their finding is confirmed by the international scientific community. And yet, society fails to take it seriously and does not muster the will to avert the catastrophe. Director Adam McKay intended the film to be a metaphor for inaction in the face of climate change.

Is it a good analogy? Maybe. In the movie, we see a society that is distracted by endless entertainment, obsessed with consumerism, polarized by political factions, distrustful of science, lashed by the imperatives of profit, riven by inequalities, and fractured into alternate media realities. Sadly, there is much truth in this picture. Let's look at some of the strong analogies.

1. *Polarization and alternate realities.* The rich and powerful who profit from the status quo certainly do not always act in good-faith ways to find common interest climate policies. In the film, the political party in power coins the slogan "Don't Look Up." They convince their supporters that the scientists are just trying to scare them. Partisan affiliations run deep in people's sense of self and belonging, strongly shaping one's mental models. And alternative media (even alternative "constitutions of knowledge" as we will see) allow for the development of incommensurable views of reality. Confirmation bias happens when people selectively uptake only the information that readily fits into their mental models and group identity. We often see only what we want to see. In the film, eventually even the most determined denialists are forced to acknowledge the reality of the comet, but by then it is too late.

2. *It doesn't feel real.* In the film, even the scientists find it hard to truly absorb the weight of the coming disaster. One of them keeps wondering if this is really real. At first, the comet is only known through scientific instruments and its trajectory is only known as a matter of calculations. It is all so abstract and far away. It just doesn't *feel* real even to those who *know* that it is real! Similarly, climate change is hard to believe even when you fully acknowledge all the scientific evidence. The slow-motion,

ambiguous, all-encompassing nature of climate change tends to sap our sense of urgency and, thus, our will to think through these complexities. You might even say that our brains are not evolved to handle the slow, massive, and complex realities of climate change (see Duhaime 2022).

3. *Confusion about what to do.* Even for those who take the comet seriously, there are confusions and disagreements about who, exactly, has which responsibilities to take what action. Should the US act unilaterally ... but what if they fail? Should an international collaboration be formed ... but what if that takes too much time? What role should the private sector play? These are all questions that find their analogies in climate politics. .

But wait. Consider some problems with the comet-climate change analogy.

1. *It's too precise and singular.* There is no exact moment of climate dooms-day when all of civilization will be snuffed out. Rather, there is a relatively slow accumulation of changes that will be felt differently across the globe.

2. *It's too certain.* There is not such a high-level of certainty about what the future impacts of climate change will be. This will depend on thousands of decisions, whereas the comet has an established, predictable trajectory and outcome.

3. *There's no one obvious fix.* In the film, although the implementation is fuzzy, it's pretty clear that there is really just one thing to be done: deflect the comet before it gets to Earth. In the case of climate change, though, there is a bewildering array of options on the table, all with costs, benefits, tradeoffs, and uncertainties.

4. *The ethics are too simple.* The comet is an unmitigated bad thing (well, ok, billionaires want to mine its rare elements!), but climate change is caused by people pursuing better lives via polluting systems that none-theless provide benefits. This is the primary reason why disentangling from current patterns of development is so difficult. The slogan in the film of "just look up" implies that climate change is as simple as acknowl-edging the basic fact of a comet hurtling toward Earth. It suggests that once we just acknowledge the reality of anthropogenic climate change,

the right thing to do will be self-evident. But, as we noted about the "false binary" pitfall, it is just not that simple.

The reporter Amy Harder (2017) argues that climate change is more like diabetes for the planet than a comet. Like climate change, diabetes does not cause other health conditions, but it can make them worse. Climate change and diabetes both become a bigger problem the longer we fail to address it. Yet both are not really problems that we can solve completely. Rather, they are conditions requiring life-long efforts to manage within the context of other, sometimes, competing priorities.

Then again, maybe the point of the film *Don't Look Up* is that we need to convince ourselves (like a noble lie) that climate change really is as evident and urgent as a comet hurtling toward us. Maybe the film is telling us to put ourselves in that constant state of emergency so that it stays at the top of our list of priorities.

Climate Change as Hyperobject

Now let's move from art into theory and the humanities to do more analogical thinking. The philosopher and ecologist Timothy Morton (2013) says that climate change is like capitalism, a pandemic, the internet, tectonic plates, or even "all plastic ever manufactured." What do they all have in common ... what makes them *alike*?

For Morton, they are examples of *hyperobjects*. They lack clear boundaries and centers, we cannot step outside of them, it is impossible to know everything they are "doing," and they are "massively distributed in time and space." Each bit of plastic or molecule of CO_2, for example, is just one local manifestation of a larger whole that is dispersed all over, that is weaving in and out of living and non-living systems, and that will persist for a long time relative to our human scale of perception. It is, in other words, "hyper." You can't experience hyperobjects but you also can't *not* experience them.

Morton counsels us to develop a "spirituality of care" toward all the objects of the world, including ones that are hyper, even sinister. He thinks we are too careless. We don't take reality seriously. The scholar of Indigenous knowledge Daniel Wildcat (2009) similarly talks about "indigenous realism." Reality is not just facts and figures, but stories and relationships. We are not isolated individuals and humans are not the center of creation.

We are, rather, a part of the web of life, a condition that calls for respect and care. It does seem like climate change is the result of a great deal of careless action, or at least action that could be more careful by taking into account all of the consequences of what we are doing.

Here is one way to use hyperobject theory. Climate change is:

- **Non-local**: it is everywhere and nowhere like a hemorrhage infused across all tissues. We can't touch it, but it is always touching us. It is not the weather, but it is "all over" the weather. It's hard to pin it down. We point to charts and models and then bush fires and heatwaves, but climate change is never like, say, a comet that we can point to and tell the skeptics to just look. Where is climate change? What are we looking at?

- **Viscous**: it sticks to everything. Climate change is in finance, infra-structure, healthcare, national defense, agriculture, immigration, energy, city planning, etc. Decisions about climate change are not always about climate change. Maybe they are about competition with China. Climate change acts like a "force multiplier" complexifying things. It doesn't cause anything on its own, but it shapes and inten-sifies things. For example, the IPCC finds with high confidence that anthropogenic climate change is leading to increased fire weather (warmer, drier conditions) in North America. So, it's too simple to say that climate change "caused" fires in Canada, but it is making them more likely and more intense.

- **Molten**: it is so massive that it metaphorically warps the fabric of spacetime. Our high-energy way of life reaches deep into the future and across the planet. This melts our old moral and political catego-ries, making us Earthlings in new ways. Do we have obligations to future generations and to people in other nations halfway around the globe? Who is innocent and who is guilty? Who is responsible when it seems like everyone and no one is to blame?

- **Interobjective**: it exists in the relations between objects. The climate system is a flux of interactions between living and non-liv-ing things. This challenges our pervasive sense of self as individual and of humanity as separate from nature. Our actions are braided into natural systems that exert their own agency. Attribution science works in this space as it attempts to tease out just how much of a storm or a forest fire was nature and how much was human. Should we sue corporations for part of the damages of all extreme weather events? They—and all of us—are there in the forces of "nature."

* **Phasing**: climate change comes in and out of our awareness. At times, like reading this book, it occupies the foreground of our consciousness, but most of the time it slips out of our view. In a way, we simply forget about it. So, even though it is omni-present, it usually phases out of awareness. This has important implications for ethics (decisions often get made without factoring in climate change) and policy (climate change tends to slip down the list of more pressing priorities that don't phase out of awareness).

Conclusion

The climate is mediated through science and other aspects of culture. It can be defined as long-term average weather or as the interaction of the Earth's systems. Human GHG emissions are causing the climate to change—we are changing the way things change. This is an urgent problem and yet it often doesn't feel urgent to us. So, do we need to develop a new kind of moral sensibility in order to act appropriately? Or can we handle the problems with our existing moral and political norms? Because we understand reality through language, we will need not just science but also art, stories, politics, and film to comprehend climate change. In particular, we'll need to develop the interpretive critical thinking skill of analogical reasoning: what is climate change like? Is it like a comet or like diabetes or the internet...? Just what *is* it?

Activities and Questions

1. How would you describe the weather-ways where you live? Have you noticed them changing? What do IPCC reports say about climate in your region? Do your elders talk about changes? How does this make you feel?

2. Watch *Don't Look Up* and discuss in class. What lessons do you take from it? Is it a good analogy for climate change?

3. What does a "spirituality of care" mean? Do you think it is central to dealing with climate change? A great resource here is the novel *Parable of the Sower* by Octavia Butler.

4. How can you represent climate change via art? Can you capture it in a drawing, a film, a story, a dance, etc.? In what ways is art like and not like science?

References

Allen, M.R., et al. 2018. "Framing and Context." In *Global Warming of 1.5°C: An IPCC Special Report*, 49–92. Cambridge: Cambridge University Press.

Duhaime, Ann-Christine. 2022. *Minding the Climate: How Neuroscience Can Help Solve Our Environmental Crisis*. Cambridge, MA: Harvard University Press.

Hansen, James, Makiko Sato, and Reto Ruedy. 2023. "Uh-oh. Now What? Are We Acquiring the Data to Understand the Situation?" Communications post on Dr. Hansen's website, August 14. https://www.columbia.edu/~jeh1/mailings/.

Harder, Amy. 2017. "Climate Change Is Like Diabetes for the Planet." *Axios*, September 11.

Hayhoe, Katherine. 2019. "What's the Big Deal with a Few Degrees?" Climate Weirding. March 20. https://www.youtube.com/watch?v=6cRCbgTA_78.

Hulme, Mike. 2017. *Weathered: Cultures of Climate*. London: Sage.

IPCC. 2021. *Climate Change 2021: The Physical Science Basis. Contribution of Working Group I to the Sixth Assessment Report*. Cambridge: Cambridge University Press.

—. 2022. "Summary for Policymakers." In *Climate Change 2022: Impacts, Adaptation, and Vulnerability. Contribution of Working Group II to the Sixth Assessment Report of the Intergovernmental Panel on Climate Change*. Cambridge: Cambridge University Press.

Morton, Timothy. 2013. *Hyperobjects: Philosophy and Ecology After the End of the World*. Minneapolis: University of Minnesota Press.

Overpeck, Jonathan, and Bradley Udall. 2020. "Climate Change and the Aridification of North America." *Proceedings of the National Academy of Sciences* 117 (22): 11856–58.

United Nations. 2022. *Global Assessment Report on Disaster Risk Reduction*. Geneva: UN Office for Disaster Risk Reduction.

Wallace-Wells, David. 2022. "Beyond Catastrophe: A New Climate Reality Is Coming into View." *New York Times*, October 26. https://www.nytimes.com/interactive/2022/10/26/magazine/climate-change-warming-world.html.

Wildcat, Daniel. 2009. *Red Alert! Saving the Planet with Indigenous Knowledge.* Golden, CO: Fulcrum.

CHAPTER 3

Framing the Climate Problem

What Are the Goals?

Global Greenhouse Gas Emissions and Warming Scenarios

- Each pathway comes with uncertainty, marked by the shading from low to high emissions under each scenario.
- Warming refers to the expected global temperature rise by 2100, relative to pre-industrial temperatures.

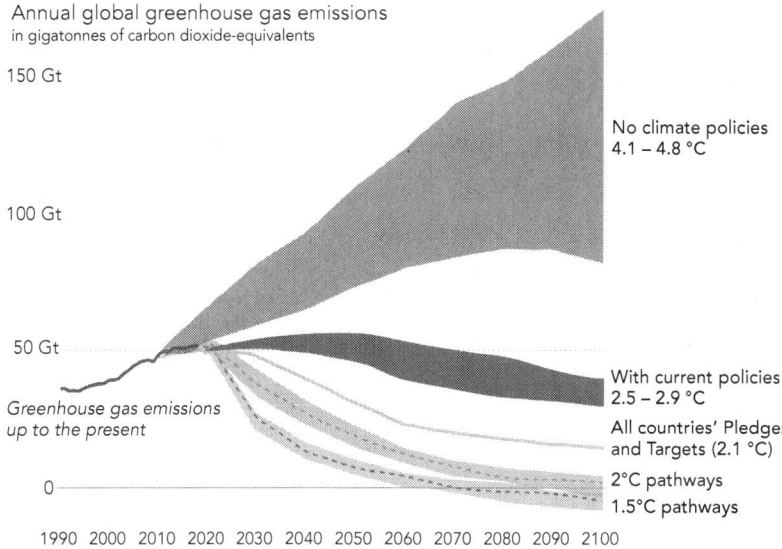

Annual global greenhouse gas emissions
in gigatonnes of carbon dioxide-equivalents

FIGURE 3.1

I n what ways is anthropogenic climate change a problem? Maybe it is a whole bunch of problems. What would it mean to solve or fix the climate problem?

"Framing a problem" is the most important aspect of our critical thinking skills. It is all about meaning-making. Put differently, it is another instance of "what is" questions. When people argue about problems, it's like they are seeing different realities. When a developer, for example, wants

65

to build homes on their property, they talk about jobs, tax revenue, and economic growth. The people in the established neighborhood nearby, however, talk about traffic, noise, and quality of life. Someone will say, "this is really about x." Another person will say, "this is all about y." What *is* the problem? How is it decided? Who gets to decide?

When it comes to climate change, the physical sciences play an important role in framing the problem, because they are needed for accessing the reality of the climate. Modern science does so much. It gives us a coherent and detailed view of the world. It corrects pre-scientific views (about, say, the causes of climate change). Science discloses previously undetected phenomena (like the greenhouse effect or the ozone layer). And scientific laws explain how the world works (how organisms develop, how planets move, how mountains form, how diseases spread, etc.).

Yet science is insufficient for problem-oriented thinking. A problem is not something the sciences will ever detect, no matter how sensitive their instruments are. Problems are not out there in the physical furniture of the universe. They are framed, constructed, and contested. And scientific explanation only gets underway when it is clear what problem needs to be explored. Why do we care about the climate? What causes us to bring our attention to it?

It is these perspectives of care, our values, that direct our attention. They tell the sciences what to look for and they shape our interpretation of scientific findings. The sciences can't tell us what values to pursue or how to weigh competing goals. Problems entail reasoning about values; they are how ethics and politics get tangled up with the climate sciences. Problem-framing is how we organize these tangles so that we can make sense of climate change and guide our efforts to address it.

So, to have a problem, we need to have values that are not being met and might not be met in the future unless we change our ways. Indeed, problem-framing is always forward-looking, because it is about what we should do to make things better in the future. As Figure 3.1 shows, the future is uncertain and open-ended. Whether we achieve our climate goals depends partly on things we can control and partly on things we cannot.

This chapter introduces the climate literacy skills of problem-framing, laying a foundation that we will build upon for the rest of the book. We'll first practice using a tool called "the problem orientation" that comes from the policy sciences (see Clark 2002). Then, we'll use this tool to examine how the United Nations Framework Convention on Climate Change (UNFCCC) frames the problem. Here, we'll focus mainly on the goals.

With its problem framework, the UNFCCC establishes the predominant discourse, informing nearly everything that is debated and done. We'll conclude by complementing this political framework with two theories: wicked problems and collective action problems. These theoretical lenses give us further insights into how people frame climate change as a problem.

The Problem Orientation

To begin using this tool, let's consider a stylized personal problem and then a public one.

EXAMPLE 1: A Personal Problem

Fernando steps on the scale: he weighs 200 lb. That's not a problem, it's a number. A week ago, he weighed 190 lb, and two weeks ago he weighed 180 lb. That's still not a problem, it's a trend: 180, 190, 200. But Fernando wants to lose weight. Now we have the bare minimum for a problem: a *goal* that is not aligned with *trends*. To address this problem, he will need to consider the *causes* of the trends and his weight *projections* in the future. Perhaps most importantly, he'll need to think about the available *alternatives*: what can he do differently to accomplish his goal? This is the problem orientation framework of GTCPA:

* **Goals**: lose weight.
* **Trends**: gaining weight.
* **Causes**: too many doughnuts.
* **Projections**: increased weight gain.
* **Alternatives**: stop eating doughnuts.

Now, anyone who has thought about their weight will say that this is too simple. Let's take the goal for starters. Does Fernando have an ideal weight that could be a metric for the goal? Is weight loss the only goal? Is it the best way to characterize his goal? Maybe something like "improved physical health" is a better formulation. Weight is just one metric, after all. It is possible to get healthier and actually gain weight, say, by adding muscle mass. Most importantly, goals often conflict. Doesn't Fernando also want to enjoy delicious foods? I mean, doughnuts are good! Are goals in conflict? How to balance them?

Next, let's turn to causes. It is rarely the case that a single cause is responsible for a trend (note how talk of causality is about responsibility). What else might explain (cause is also about explanation and diagnosis) Fernando's weight gain? Did he stop exercising? How is he sleeping? What else is he eating? So, multi-causality is almost always happening. When it comes to climate change, this is even more the case. Forest fires, for example, can be caused by lightning strikes, campfires, fire-suppression policies, development patterns, drought, invasive beetles, pyromaniacs, *and* warming trends.

Sometimes it is helpful to talk about causes that are more direct and those that are indirect. Why is Fernando eating so many doughnuts? Or you might say: what is causing this cause? Maybe Fernando has underlying metabolic issues going on, which themselves may be a complex mix of environmental and genetic causes. Or maybe a grocery store nearby went out of business, leaving him with fewer food options. Or maybe he got exposed to new doughnut ads or moved in with a new doughnut-loving roommate.

For these reasons, problem-oriented thinkers often talk about "conditioning factors" rather than "causes." This is a way to acknowledge multiple, layered, often indirect factors that shape a situation even if they are not straightforwardly *the* cause. In climate policy, people similarly talk of direct and indirect drivers of change. This makes discussions about responsibility and explanation more complex. No one straightforwardly causes climate change, which in turn does not straightforwardly cause any particular weather event. It is a tangle of conditioning factors.

Figure 3.1 illustrates projections, and we'll talk more about them later. For this example, let's turn to alternatives. Clearly, if there are multiple goals at stake and many conditioning factors at work, then the available alternatives will also be diverse and complex. That raises questions about which mix to choose. Fernando could moderate his doughnut consumption (why stop entirely?) while exercising more and sleeping better. Pursuing alternatives is all about action, which entails *will*. What is it going to take to get Fernando to adopt more healthy behaviors? Is it really that he doesn't know that doughnuts are unhealthy? Or is he suffering from weakness of will? Many smokers *know* it is unhealthy and *want* to quit, but they just can't. Is climate change like that?

EXAMPLE 2: A Public Problem

Things get really complicated when we move into the public sphere, because now there are multiple stakeholders that might each have their own way of framing the problem.

I once took part in a policy process about electricity in my home city of Denton, Texas. We have a utility called Denton Municipal Electric (DME), which serves all of our homes and businesses. DME is guided by three goals: (a) provide reliable electricity that is (b) affordable and (c) sustainable.

DME was doing pretty well on the first two goals. However, our community had expressed a collective desire to move rapidly toward 100 percent renewable electricity. At the time, DME was supplying electricity from the Texas grid, which was about 70 percent fossil-fuel generated. Renewables like wind and solar were displacing some of this fossil energy, but not rapidly enough for our ambitions. DME came up with an alternative: we could buy electricity via power purchase agreements (PPAs) directly from wind and solar generators. That is, we could avoid buying from the dirtier grid by entering into contracts with specific renewable energy producers.

So, here was the way DME thought about the problem:

- **Goals**: (a) reliability, (b) affordability, (c) sustainability.
- **Trends**: (a) aligned, (b) aligned, (c) out of alignment—too much fossil-fuel generated electricity and transition to renewables is not quick enough.
- **Conditioning Factors**: for (c) = buying electricity from a grid that is dominated by natural gas and coal.
- **Projections**: if we maintain status quo systems and decisions, for (c) = continued reliance on fossil fuels longer than the community desires.
- **Alternatives**: PPAs to directly buy wind and solar energy.

Sounds great, right? Well, there was a catch. Buying the PPAs carried with it financial risk. Being very cognizant of goal (b) affordability, DME didn't think they could responsibly adopt this alternative course of action without some kind of financial hedge to lower the risk of rate hikes. They decided that the only viable path would be to build a natural gas-fired electricity generator that they would own and operate. That way, under the right market conditions, DME would be able to sell electricity from the gas plant, thereby defraying the costs and mitigating the financial risks of the renewable PPAs.

This is where things got controversial. Some residents were opposed to the gas plant. A few of them were worried about the debt required to build the plant, but most had climate-motivated reasons: they insisted that climate action means no new fossil fuel infrastructure. Period. Not under any circumstances could this be morally justified. Others, however, noted that any electricity sold from the new natural gas-fired power plant would displace dirtier electricity on the grid. Plus, it was a financially secure way to make huge strides toward our 100 percent renewable electricity goals. The opposition, however, asked how a city could claim to be 100 percent renewable while operating fossil infrastructure. What do the goals of 100 percent renewable and sustainability really mean in this case?

The politics got heated. Some friendships fell apart, even though people shared the same three goals! Questions were raised about whether DME could be trusted. An independent consulting agency was hired to review the proposed plan. Competing analyses got complex as different assumptions and methods generated different outcomes. Concerned residents kept asking about other alternatives. Could we, for example, purchase grid-scale batteries to store electricity rather than build the gas plant? DME concluded that would be cost prohibitive, but others disagreed. Who were the real experts? What's the right decision?

The Predominant Climate Change Problem Framework

That's just a gloss of one small instance of climate politics to show how the GTCPA framework can organize our thoughts, but also how there is no formula to make hard decisions easy. What we need to see now is how the Denton case—like all of climate politics—is situated within a broader global discourse that is anchored in a predominant problem framing.

Adopted in Rio de Janeiro in 1992 by 154 nations, the UNFCCC coordinates the political, social, and economic responses to climate change on an international level. It acts like a bridge between the IPCC's scientific assessments and international political treaties such as the 2015 Paris Agreement. The UNFCCC's framing of the problem largely establishes the parameters for climate politics and policy from the local level to the international level.

Let's use the problem orientation to break it down. In this chapter, our focus will be on the goals—what they are and how they are measured and pursued. The remainder of the book will cover data on trends and projections and survey debates about conditioning factors and alternatives.

The UNFCCC states that the goal is to achieve:

stabilization of greenhouse gas concentrations in the atmosphere at a level that would prevent dangerous anthropogenic interference with the climate system. Such a level should be achieved within a timeframe sufficient to allow ecosystems to adapt naturally to climate change, to ensure that food production is not threatened and to enable economic development to proceed in a sustainable manner. (UNFCCC, Art. 2, 1992)

There are basically *two big goals*, safety and well-being:

1. Safety: reduce the risks and impacts of anthropogenic climate change
   a. Mitigation: prevent and minimize dangers by reducing GHG emissions (and by increasing carbon sinks)
   b. Adaptation: increase security and resilience to withstand climate dangers
2. Well-being: continue with economic development

Note the vital point: *Goal 2 has long been in opposition to Goal 1.* Indeed, with its GHG emissions, development (2) is the key conditioning factor of climate dangers (1).

As the economist Robert Mendelsohn puts it, the challenge is creating a future with a nineteenth-century carbon footprint without backsliding into nineteenth-century standards of living (in Shannon 2022). Can you picture what this might look like? Are there models of this kind of life in existence now? Note that this framing of "nineteenth-century standards of living" assumes a developed world perspective—what about people who still live at or below those standards?

The UNFCCC says that we want to improve the standards of living for billions of people but to do so in ways that are climate safe. Up to this point, gains in human development have largely come from fossil fuels and industrial agriculture, that is, activities that emit GHGs. In the future, this predominant problem framing says that we want to keep development going, but with other energy sources and technological means such as solar panels, wind turbines, and electric vehicles. We'll see the key term here is *decoupling*: to have both goals, we need kinds of development that are no longer linked to dangerous climate risks and impacts.

If we didn't care about development, we could just decommission power plants and pipelines. Yet, the economy would tank, causing mass suffering. If we didn't care about climate risks, we could just keep going with status quo energy and agricultural practices. Yet, extreme weather events would multiply, causing mass suffering. We need to attend to both goals. That's why climate politics is hard. People, for example, can't stop using natural gas appliances in their homes overnight. Yet we also must find ways to quickly transition. Those phrases "overnight" and "quickly" resonate with the UNFCCC's emphasis on "time-frame." What is the right pace of climate action? This also factors into debates between those "concerned" about *climate change* and those "alarmed" about the *climate crisis* like we saw in the Introduction.

Measuring the Goals

A problem occurs when trends (and/or projections) and goals are misaligned. To know whether that is the case, we need to specify the goals in ways that are measurable. That gives the sciences the guidance they need to look for the relevant trend data. As we'll see, this is fraught, but first let's briefly look at each goal in turn.

GOAL 1A. MITIGATION: Prevent and Minimize Dangerous Interference in the Climate System

What counts as "dangerous interference"? The UNFCCC has measured this in terms of temperature targets. Article 2 of the Paris Agreement states that the goal is to hold the increase in global average temperature "to well below 2°C above pre-industrial levels." In the following years, the international consensus has largely settled on 1.5°C as the threshold for "dangerous interference." The 1.5°C goal is highly ambitious; frankly, it is unrealistic. As emissions continued to rise after the Paris Agreement, it became clear that we are almost certainly going to exceed the 1.5°C goal (see Hansen et al. 2023). This does not mean that civilization will suddenly collapse when we hit 1.51°C. Rather, it means that each additional increment of warming is more and more dangerous: 1.6°C is better than 1.7°C and so on.

From the pre-industrial period to 2024, global mean temperature already increased 1.2°C. This is at the heart of the climate problem: temperature is trending toward the danger zone.

The safety mitigation goal (1a) is also quantified in two other important ways.

* *Net Zero*. The goal by 2050 is to negate GHGs through emissions reductions and carbon capture. This is what it would take to stay under the 1.5°C threshold: not just eliminating emissions but also pulling carbon out of the atmosphere. The term "net zero" means that any remaining human contributions of GHGs are negated by human activities that remove the same amount of GHGs from the atmosphere. Effectively, we will need to increase carbon sinks. It's likely that some sectors of the economy will be very hard to decarbonize, so it looks increasingly important to develop carbon capture technologies in order to reach net zero (see Chapter 12). This is especially true given the fact that total GHG emissions continued to climb long past the Paris Agreement.

* *Carbon Budget*. Since temperature increase is driven in large part by carbon emissions, you can think of a total amount of carbon that can be burned before crossing the 2°C threshold. Figure 3.2 shows the carbon budget from 2013 for a 50 percent chance of remaining under 2°C of warming. It shows that most fossil fuel reserves have to stay in the ground (unless we get really good at pulling carbon out of the atmosphere). As emissions continue to climb, the budget shrinks.

There are numerous other metrics for "dangerous interference" including sea-level rise, ocean acidification, and extreme weather events like heatwaves. These are all vital to monitor, and many are trending in the wrong direction. The point here is that on a global level, all of these metrics are conditioned by temperature increase, which is conditioned by GHG emissions.

Global Carbon Budget for a Two-Degree World

The carbon budget refers to the maximum quantity of carbon (in billion tonnes) that can be released to maintain a 50 percent probability of global average temperature rise below 2°C (the target set by the UN Paris climate agreement). This has been measured relative to the quantity of carbon released if all fossil fuel reserves were burned without using carbon capture and storage (CCS) technology. The difference between the two is defined as 'unburnable carbon'.

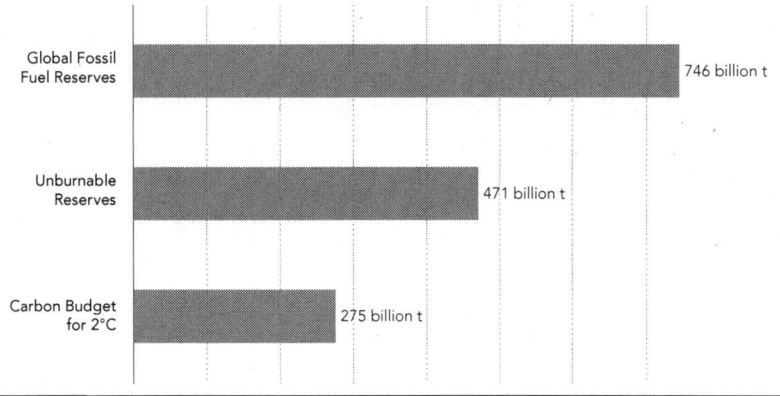

Global Fossil Fuel Reserves — 746 billion t

Unburnable Reserves — 471 billion t

Carbon Budget for 2°C — 275 billion t

FIGURE 3.2

GOAL 1B. Adaptation and Resilience

It is too late to prevent some impacts from anthropogenic climate change. Plus, climate dangers will always exist, even in the absence of anthropogenic GHG emissions. Thus, another major goal is to get stronger in the face of danger. The IPCC defines adaptation as "the process of adjustment to actual or expected climate and its effects." And it defines resilience as "the capacity of social, economic, and environmental systems to cope with a hazardous event or trend or disturbance." When resilience increases, vulnerability decreases.

More resilient systems have a higher threshold for danger, because they are more capable of functioning under duress. After all, what we really care about are the abilities of systems (living or artificial) to function or flourish. Indeed, whether severe weather can be attributed to anthropogenic climate change or not, it is still a commonsense idea to reduce vulnerability. Implicit in this goal is also diversity, because extinction is what happens when things are unable to adapt, and because diversity tends to increase resilience for systems.

That we are dealing with diverse systems means that the metrics and trends here will largely be context dependent. Indeed, the IPCC structures its reports on adaptation and impacts around several key systems, including terrestrial ecosystems, cities, healthcare, oceans, agriculture, and water. We might, for example, examine trends in the health of coral reefs and associated marine biomes.[1] Or we could look how individual species or even entire forests are migrating with increasing temperatures. For human systems, we can examine trends in the development of and equitable access to vulnerability-reducing infrastructure[2] and services.

The ongoing development project can make this confusing, because it is putting more people, wealth, and infrastructure in harm's way. Imagine, for example, that the exact same hurricane (size, strength, speed, etc.) crosses Florida in 1944 and again in 2022. The second one will cause significantly more damage, not because of climate change (in this thought experiment, they are the same storms), but simply because there is more stuff in the storm's path—bigger cities, more roads, etc. One way to account for this is to "normalize" trends in damage through methods that model how costly storms in the past would have been had they occurred under present development conditions (see Weinkle et al. 2018). Resilient and adaptive development, then, means building human systems capable of withstanding extreme weather events. It means that, over time, human well-being (however defined!) increases even in the face of hardships.

And there is evidence that this is happening. Consider, for example, Figure 3.3, which shows deaths from all types of natural disasters. Development has generally made people safer from climate and weather extremes (with some important exceptions for those who have been displaced, disposed, and marginalized[3] by development). The big question, of course, is whether this trend will continue or whether increasing climate dangers will reverse it.

GOAL 2. Well-Being

As noted in Chapter 1, Gross Domestic Product (GDP) is the common metric used for economic development and its underlying goal to improve human well-being. This has certainly been on an increasing trend, though *crucially* standards of living are far from equitably distrib-

---

1   The plant and animal community of a major climatic region or type of habitat.

2   The permanent subordinate physical or structural parts of something—in this case, the parts necessary for the functioning of a community: e.g., water ducts, waste disposal facilities, roads, buildings, power supplies.

3   Treated as insignificant, or ignored.

Global Deaths from Disasters over More Than a Century

The size of the bubble represents the estimated annual death toll. The largest years are labeled with this total figure.

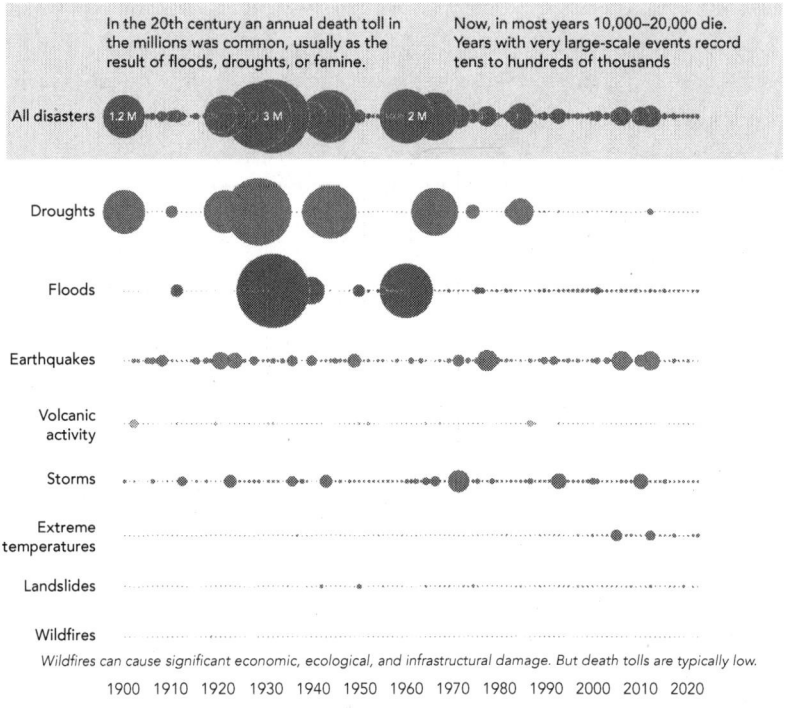

In the 20th century an annual death toll in the millions was common, usually as the result of floods, droughts, or famine.

Now, in most years 10,000–20,000 die. Years with very large-scale events record tens to hundreds of thousands

All disasters   1.2 M   3 M   2 M

Droughts

Floods

Earthquakes

Volcanic activity

Storms

Extreme temperatures

Landslides

Wildfires

Wildfires can cause significant economic, ecological, and infrastructural damage. But death tolls are typically low.

1900 1910 1920 1930 1940 1950 1960 1970 1980 1990 2000 2010 2020

FIGURE 3.3   Data Source: EM-DAT, CRED/UCLouvain Brussels, Belgium, www.emdat.be (D. Guha-Sapir), OurWorldInData.org

uted across the globe (see Figure 3.4). We also noted that perhaps "sustainable" development means using metrics beside GDP growth to track progress. The UN Sustainable Development Goals (SDGs) account for a wider range of metrics. Maybe we should be tracking happiness, sharing, enlightenment, or human harmony with nature. After all, happiness does not continue trending upward after a certain level of GDP per capita has been secured. Perhaps the development that matters is spiritual or about one's character. All of this is harder to measure, of course. Still, though, it gets you wondering: what are the trends that really matter when it comes to human development or progress? What does the goal of well-being really *mean*?

We noted above how this goal (2) has been in opposition to goal 1a. Development means GHG emissions, which mean danger. But that's not

GDP Per Capita, 1650 to 2018

This data is adjusted for inflation and for differences in the cost of living between countries.

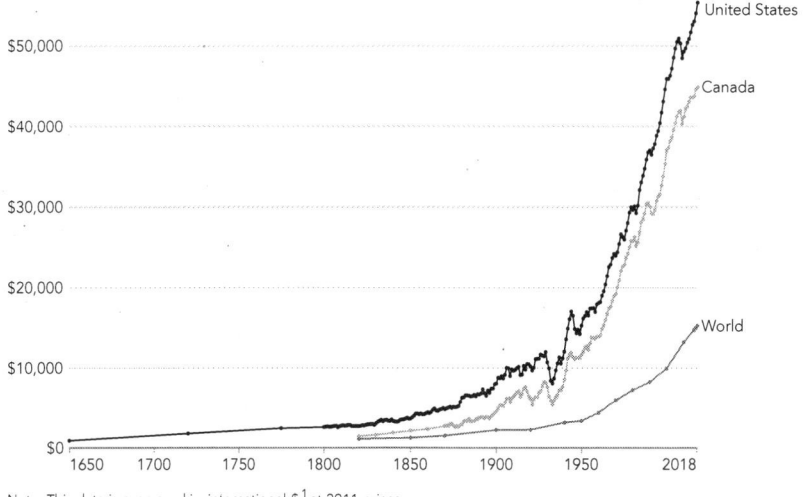

Note: This data is expressed in international-$ [1] at 2011 prices.

1. International dollars: International dollars are a hypothetical currency that is used to make meaningful comparisons of monetary indicators of living standards. Figures expressed in international dollars are adjusted for inflation within countries over time, and for differences in the cost of living between countries. The goal of such adjustments is to provide a unit whose purchasing power is held fixed over time and across countries, such that one international dollar can buy the same quantity and quality of goods and services no matter where or when it is spent. Read more in our article: What are Purchasing Power Parity adjustments and why do we need them?

FIGURE 3.4

the whole story, because development also can mean resilience, goal 1b. People with access to infrastructure and services are better equipped to handle climate extremes. As noted, natural disaster costs increase every year, because the technosphere[4] is growing due to development, thus, more valuables are exposed to harm. However, the costs as a percentage of global GDP have actually been decreasing, which is one way to say that the total system is more resilient: it can withstand losses at a lower cost to the whole (see Pielke 2019). There is truth in the saying that wealthier is safer.

So, the goal of development is manifold and it is in a complex relationship with the goal of safety. For example, consider how the development of new technologies like genetic engineering can help adapt our food systems to a changing climate. Development both causes climate change (exposing us to new risks) and shelters us from the impacts of climate change (reducing vulnerability). This is an important tension to keep in mind. Finally,

---

4   That part of the environment modified by human action.

note that the goal of development is open-ended in a way that the temperature target is not. Do we ever have enough, or too much, development? We'll take this up in the next chapter.

Pursuing the Goals: Justice and Finance

We've seen how not all people are equally responsible (conditioning factors) for climate dangers. The developed world has generated the vast majority of emissions and the developing world is generally most vulnerable to the resulting dangers. This basic asymmetry is the unfairness at the heart of climate justice.

This is why the UNFCCC notes in Article 2 of the Paris Agreement that climate decisions should be "implemented to reflect equity and the principle of common but differentiated responsibilities and respective capabilities." Everyone should contribute to solving the problem in ways proportionate to their responsibilities and capabilities. We all live on the same planet in common and, thus, we all have responsibilities, but those who have contributed most to the problem (created the most GHG emissions) have the greatest moral responsibility and have the greatest capabilities to help. In metaphorical terms, we are not all in the same boat—some have yachts and others only have a plank of wood to hold onto. Yet we are all in the same storm. Of course, in this metaphor, we have to also imagine that those with the yachts are making the storm worse!

Climate change responsibility often boils down to money: the polluter should pay. That's why the same article states that achieving the goals of mitigation and adaptation will require "making finance flows consistent with a pathway towards low greenhouse gas emissions and climate-resilient development." We'll return to this in later chapters on climate ethics and policy with regard to loss and damage, climate justice, and climate reparations.

It's important to emphasize here the ethics and justice dimensions of defining climate goals with simple temperature targets like 1.5°C. Consider three key issues:

- First, we are almost certainly going to exceed the 1.5°C target well before 2100 (Brahic 2022). Should scientists and journalists even openly state this honest truth? What will it mean to miss such a symbolic target? Will people understand that "dangerous climate interference" never really could be defined and measured as a specific number and continue to work toward preventing 1.51°C or 1.52°C? Will people be discouraged? Will the UNFCC suffer damage to its credibility if we pass the goal but the world doesn't collapse? Or maybe this will spur more emphasis on important adaptation measures.

- Second, targets can be meaningless if specific measures are not put in place. Setting targets does not, by itself, reduce emissions. There have to be policies in place that are politically acceptable and cost-effective. Without those, all the temperature targets and climate pledges are just talk and empty promises.

- Third, and related, a focus on simple end targets distracts from crucial questions about the means, that is, about the pathways to the goal and who is included on those paths. This limits definitions of "success" in problematic ways: there is far more to a thriving future than one temperature target. How we get to our goals matters a great deal—we need to employ means that are fair and compassionate. This is why the focus on equity and justice is vital. "Success" can't mean getting to 1.5°C in ways that leave millions of people in poverty (Hulme 2020).

Consider the unprecedented Canadian wildfires in 2023. Certainly, aggressive mitigation efforts to limit warming will help reduce the risks of extreme fire weather. But so much else is needed on the ground, including firefighting resources and forest management changes. Given that climate change is tangled up with so much, like wildfires, "success" will mean lots of things with many targets.

Climate Change as a Wicked and Collective Action Problem

This predominant problem framing of safety and development establishes the structure of political discourse. It also indicates why climate politics is so messy—because there are multiple goals that are open to different interpretations by a variety of stakeholders using an abundance of trend data from many sources. We can gain a better understanding of this messiness with the aid of a couple of theories.

Climate Change as a Wicked Problem

Design theorists Horst Rittel and Melvin Webber coined the term "wicked problem" to characterize complex social policy problems. Unlike "tame" problems, wicked problems don't have a straightforward technical or scientific solution. They involve more than just, say, straightening a bent pipe or replacing a damaged part.

Wicked problems cannot be described in linear terms, because the conditioning factors and their symptoms are vast and entangled. Complexity allows for multiple ways of defining the problem (what are the goals, what vision of the future are we pursuing?) and locating the problem (where, in the vast causal networks, does the problem lie?). Given this, Rittel and Webber conclude, "the problem of identifying the actions that might effectively narrow the gap between what-is and what-ought-to-be" becomes very difficult (1973, 159).

For a wicked problem, part of the problem is knowing what the problem even *is*! To make the problem even more wicked, interventions in systems (social or natural) will have unintended consequences, the people causing the problem have a vested interest in manipulating its formulation to perpetuate the status quo (do you think oil companies want to leave all their profits in the ground?!), and the stakes are extremely high.

Indeed, others have argued that climate change is a "super wicked problem" (see Levin et al. 2012). They argue that in addition to everything noted above:

- Time is running out (e.g., the 1.5°C goal is fast slipping out of reach).
- Those who cause the problem also seek to provide a solution (fossil fuel companies, for example, increasingly claim to be driving a grand energy transition and to be "carbon neutral" via often dubious carbon credit schemes).
- The central authority needed to address the problem is weak or non-existent (the UNFCCC has no real political authority over nation states or corporations).
- Policy actions are irrationally discounting the future[5] (arguably the impacts of climate change on future generations are being vastly under-estimated).

---

5   Counting harms (and benefits) as less important as they are farther away.

I would add to this the fact that humans are just not evolutionarily adapted to think on global scales and centuries-long timeframes.

Climate Change as a Collective Action Problem

Your personal GHG emissions cause real harm, which is often described as the "social costs of carbon" though they are ecological costs too. The philosopher John Broome (2012) estimates that the monetary value of the harm you cause over a lifetime ranges between $19,000 and $65,000. These are "externalities," meaning you don't actually pay this money. They are costs you impose on other, innocent, people. Your lifetime emissions wipe out more than six months of a healthy human life.

These harms, however, are diffuse, indirect, and imperceptible. Meanwhile, the benefits you derive from the activities that produce the emissions are direct and tangible. You get the electricity, the heat, the travel, the food, etc. As you enjoy these benefits day-to-day, you can forget about the miniscule, invisible harms that come with them. Moreover, you can tell yourself that you are not causing climate change. You could eliminate your emissions entirely and it would not change the situation at all. It doesn't make sense to sacrifice for the greater good, because the sacrifice doesn't matter in the scheme of things.

So, why should you stop eating meat, forego flying, or take shorter showers ... especially when others are not doing these things? Indeed, why should the state of California or New York implement costly policies to slash GHG emissions if other states don't do the same? The entire nation of France could eliminate all their emissions and that would just be a 0.8 percent reduction globally. If it caused them economic hardship to do so, why would they?

A collective action problem is a situation in which all individuals would be better off cooperating but fail to do so. Rather, individuals pursue their own short-term gain at the expense of the group's best long-term interests. Like public radio, we all benefit from a habitable climate whether we help support it or not. Thus, there is a strong incentive to act as a "free rider," taking advantage of the public good without taking responsibility for sustaining it. Modern societies tend to prioritize individual liberty, free markets, and state sovereignty, which means they find it difficult to implement the restraints on freedom and to foster the cooperation needed to tackle collective action problems. A commitment to personal freedom and state sovereignty above all else makes it very hard to form a collective will that is strong enough to address climate change.

This is why collective action is imperative—relying solely on individual virtue is almost certainly doomed to failure. Besides, individuals are severely constrained in the choices they can make to achieve the goals of mitigation and adaptation. In a world powered by fossil fuels and fed by industrial agriculture, it is sometimes impossible to opt out just with individual choices. Other choices are sometimes only attainable to those with sufficient means to afford greener alternatives.

What kind of collective decisions to make? That is a question to take up in later chapters on the politics and policy of alternative courses of action.

Conclusion

We might say that climate change is a wicked, collective action problem that requires making the ongoing global development project "climate safe." This will take mitigation, that is, rapidly reducing our GHG emissions. And it will necessitate more adaptation and resilience. These activities should be done and paid for in ways that are consistent with the principle of "common but differentiated responsibilities." Altogether, this is "sustainable development" where the first goal of safety is what "sustainable" means and the second goal of well-being is what "development" means. Another term for it is "green growth." Can we really continue growing the economy while stewarding a safe climate? Because that is such a central question, we'll end the first section of the book by examining it.

Activities and Questions

1. Perhaps the most famous expression of the collective action problem in environmental terms is the ecologist Garret Hardin's 1968 essay "The Tragedy of the Commons." After reading that, look up the work of the economist Elinor Ostrom. How does her work enable us to critique some of Hardin's assumptions and arguments?

2. What problems related to climate change are being debated in your local community? How are they being framed by different stakeholders and

how do they fit (or not) into broader UNFCCC framing? Can you use the GTCPA framework to analyze the decision process and even offer advice to the stakeholders?

3. What are other examples of wicked problems and how can we relate the concept of a wicked problem to hyperobjects?

4. What criticisms can you offer of the way the UNFCCC has framed climate change as a problem? Read the full text of the Paris Agreement. Do you think I have adequately represented the way it defines the problem?

References

Brahic, Catherine. 2022. "The World Is Going to Miss the Totemic 1.5°C Climate Target." *The Economist*, November 5. https://www.economist.com/interactive/briefing/2022/11/05/the-world-is-going-to-miss-the-totemic-1-5c-climate-target.

Broome, John. 2012. *Climate Matters: Ethics in a Warming World.* New York: W.W. Norton.

Clark, Susan. 2002. *The Policy Process: A Practical Guide for Natural Resource Professionals.* New Haven, CT: Yale University Press.

Hansen, James, et al. 2023. "How We Know that Global Warming Is Accelerating and that the Goal of the Paris Agreement Is Dead." Communications post on Dr. Hansen's website, November 10. https://www.columbia.edu/~jeh1/mailings/.

Hulme, Mike. 2020. "Is It Too Late (to Stop Dangerous Climate Change)?" *Wires Climate Change* 11 (August): 1–7.

Levin, Kelly, et al. 2012. "Overcoming the Tragedy of Super Wicked Problems: Constraining Our Future Selves to Ameliorate Global Climate Change." *Policy Sciences* 45 (May): 123–52.

Pielke, Roger. 2019. "Surprising Good News on the Economic Costs of Disasters." *Forbes*, October 31. https://www.forbes.com/sites/rogerpielke/2019/10/31/surprising-good-news-on-the-economic-costs-of-disasters/?sh=5186c04f1952.

Rittel, Horst, and Melvin Webber. 1973. "Dilemmas in a General Theory of Planning." *Policy Sciences* 4 (June): 155–69.

Shannon, Noah Gallagher. 2022. "What Does Sustainable Living Look Like? Maybe Like Uruguay." *New York Times*, October 5. https://www.nytimes.com/2022/10/05/magazine/uruguay-renewable-energy.html.

Weinkle, Jessica, et al. 2018. "Normalized Hurricane Damage in the Continental United States 1900–2017." *Nature Sustainability* 1 (November): 808–13.

CHAPTER 4

Green Growth or Degrowth?

Decoupling CO_2 Emissions and GDP in the UK

Consumption-based emissions[1] are national emissions that have been adjusted for trade. This measures fossil fuel and industry emissions.[2] Land use change is not included.

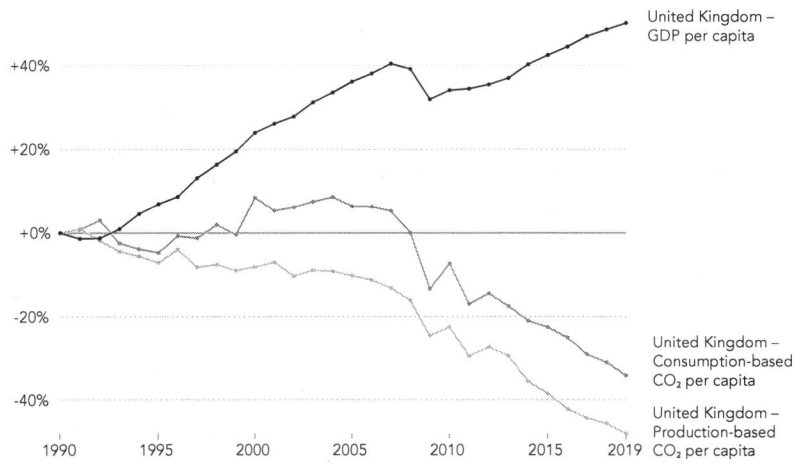

Note: GDP figures are adjusted for inflation.

1. **Consumption-based emissions:** Consumption-based emissions are national or regional emissions that have been adjusted for trade. They are calculated as domestic (or 'production-based' emissions) emissions minus the emissions generated in the production of goods and services that are exported to other countries or regions, plus emissions from the production of goods and services that are imported. Consumption-based emissions = Production-based – Exported + Imported emissions.

2. **Fossil emissions:** Fossil emissions measure the quantity of carbon dioxide (CO_2) emitted from the burning of fossil fuels, and directly from industrial processes such as cement and steel production. Fossil CO_2 includes emissions from coal, oil, gas, flaring, cement, steel, and other industrial processes. Fossil emissions do not include land use change, deforestation, soils, or vegetation.

FIGURE 4.1

Historically, economic development has been correlated with greenhouse gas emissions. Combusting carbon makes money and drives GDP growth. This needs to change, but how? The predominant climate problem framing calls for "green growth," that is, continued economic development without the GHG emissions. In climate policy, this is called *decoupling*: breaking the link between "environmental bads" and "economic goods" or achieving economic growth while reducing emissions. It can be seen in the K-shaped curves of Figure 4.1. This shows the decoupling of development from danger. A key related term is *decarbonization*: running the global economic engine on energy sources that do not emit CO_2 and other GHGs.

"Absolute decoupling" means emissions decrease as the economy grows. This is happening in many developed countries. "Relative decoupling" happens when GHG emissions decline per unit of GDP. That means that the economy is getting more efficient. The world has had relative decoupling, but overall emissions keep rising, because the economy is growing faster than it is gaining in efficiency. The big goal, then, is global, absolute decoupling: the global economy grows while emissions shrink. Some forecasts call for a doubling in global GDP between 2024 and 2050.

I think of decoupling as an enormous gamble. We are betting on technological innovation, economic dynamism, and political cooperation. This means major industrial policies to incentivize clean energy and carbon capture and generally steer billions of consumer choices down the path of decarbonization. There are some promising signs like the plummeting costs of renewable energy and the passage of landmark climate policies. And there are worrying signs like the persistent rise in GHG emissions and increase in heat waves and other severe weather events. Most climate analysts think we have no choice but to pursue green growth. Others think this is a recipe for disaster and that we must start a project of planned, rapid degrowth of the economy.

In this chapter, we'll examine this debate, which gets to the heart of climate politics, because it is about our goals and our visions of the future. Note, for example, how "development" is so often conflated with "growth," which is in turn usually defined by rather crude metrics of production and consumption. Is that what well-being—the goal of development—really means? We'll first examine three metaphors about our situation to uncover the assumptions in this debate. Next, we'll make the case for green growth— that it is both unavoidable and good. Then, we'll examine arguments against green growth—maybe green growth is neither possible nor desirable.

Three Metaphors of Sustainable Development

At the heart of our (often unthinking) commitment to green growth is a paradox: even though the planet is finite, we can indefinitely grow the economy. Can this possibly be true? Drawing from the thoughts of Rasmus Karlsson (2015), let's examine three ways of thinking about green growth.

1. The Footprint

This is the metaphor of degrowth. Maybe the growing development project (the Anthropocene) is like an expanding footprint. It is stomping over more and more of the natural world, upsetting its order and balance. It is like bacteria multiplying in a petri dish that will eventually run out of space and resources, because they outstrip the "carrying capacity" of the environment. In other words, the footprint metaphor has a subtractive logic whereby development eats away at a static fund of natural resources. Eventually, the fund will be depleted and the balance irreparably damaged. We will foul our own nest and collapse.

If this is our situation, then the answer seems clear: limits. We must reduce our impact, our footprint (see Meadows et al. 2004 for the most influential version of this argument). We need to rapidly and radically curtail consumption and production. That is, we must put an end to capitalism ... at least as we presently know it. According to this view, it is simply impossible for over eight billion people to enjoy materially affluent lifestyles. Some argue that it will be liberating to abandon a way of life based on false needs and an economy rigged against the working class. Yet it is not clear how this can be feasible. People are not eager to sacrifice modern conveniences, even if it would make them happier. Further, many of those conveniences have become genuine needs that we can no longer do without—at least not without radical structural changes to the way we grow food, build cities, work, and lead our lives. And how can we shrink our footprint—the economy—without sacrificing the security it provides? How, especially, can we do so without harming the poor and vulnerable? That would seem to require a level of solidarity and sharing—redistributing wealth—that is sadly unlikely.

2. The Rocket

This is the metaphor of green growth. Maybe development is like a rocket blasting off toward outer space where humanity can achieve a more secure cosmic foothold. This rocket keeps rebuilding itself as it gains elevation—replacing old parts (like coal plants) with new ones (like solar panels). It learns to account for problems as it travels and uses the resulting innovation to propel it ever higher, that is, to generate more and more wealth.

The footprint is a static snapshot of present technological and ecological conditions. By contrast, the rocket evokes a dynamic political imagination of development as an evolutionary process. By taking a process-oriented view with a longer time horizon, the rocket metaphor notes that achieving long-term sustainability will require passing through a temporary bottleneck of unsustainability. It is true that current conditions cannot be sustained, but they do not have to be. We will break through the current bottleneck into the "good Anthropocene" where billions live affluently on a planet where nature is restored thanks to ongoing innovation. This is our high-risk, high-reward gamble.

3. The Runway

Maybe reality is a mix of these two metaphors. What if the development project is like an airplane heading down the runway of modernity? The runway is shrouded in fog, which makes it impossible to know how long it is. There are three possibilities. First, there is enough time for the airplane to take off and become like the rocket. We could achieve high-tech sustainability—green growth. Second, the runway might be too short, but we recognize that in time and decelerate into a smaller, slower, more localized way of life based on norms of frugality and simplicity. This would be the degrowth of the footprint model.

There is a third possibility, however. Maybe the runway is too short—there just is not enough time to build up enough innovation momentum to launch a sustainable rocket. Yet we don't realize that. What if the pilots of the airplane (whoever they are!) neither commit fully to the throttle or the brakes? In that case, our future is one where we overshoot the runway and crash. This outcome seems sadly plausible to me. Why? Well, we are not fully committing to the rapid technological transitions we need to make the rocket achieve a sustainable escape velocity (new oil and gas wells are still being drilled, for example). Yet neither are we committing to a radical deceleration of the global economy.

Hopefully these metaphors help you ask the big question: How can we thrive and how can we find justice in a world of new extremes—extremes in development and danger?

Green Growth

Because the stakes are so high, it is important to critically examine the green growth narrative that dominates modern society. We all tend to lead lives guided by this story (we assume, for example, that stock markets and retirement savings will keep growing), but rarely do we stop to reflect on it. In this section, we'll consult the thoughts of those who have theorized green growth and feel optimistic about it. We might call them the "up-wingers," because they see us heading up (on rockets both metaphorical and real) to a space-faring civilization. But it is maybe better to just call them the eco-modernists, because they believe that the modern development project can learn to be ecologically sustainable. First, we'll explore the underlying logic behind this belief. Then, we'll examine the practical and moral arguments in favor of green growth.

The Logic of Green Growth

Green growth is paradoxical. Here is the heart of it: "The solution to the unintended consequences of modernity is, and has always been, more modernity" (Shellenberger and Nordhaus 2011). How can this be?! The key is to see modernity or development as a process, not a thing. It is to think like the dynamic, ever-evolving rocket, not the static footprint.

Consider an example. In the 1930s, industrialized nations started using chlorofluorocarbons (CFCs) in air conditioning units and aerosol sprays. In the 1980s, scientists discovered that CFCs were eroding stratospheric ozone, which shields life on Earth from harmful UV radiation. An ozone hole was beginning to grow over Antarctica. In 1987, the nations of the world decided via the Montreal Protocol to ban the production of ozone-depleting substances like CFCs. Ozone levels have since stabilized and a full recovery is expected by 2075.

Global Emissions of Ozone-Depleting Substances

Annual consumption of ozone-depleting substances. Emissions of each gas are given in ODP tonnes.[1]

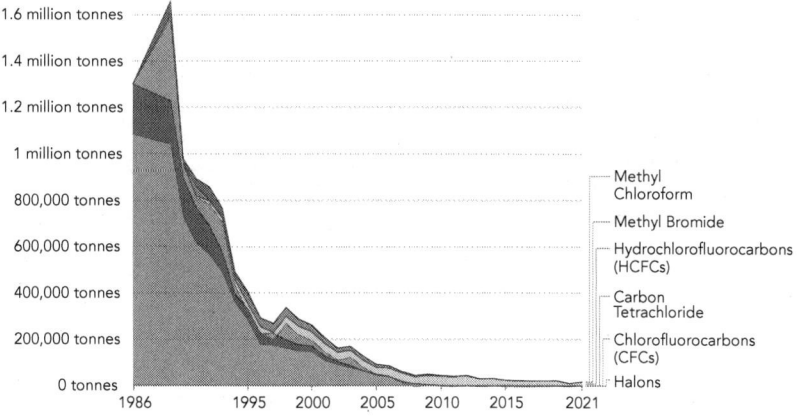

Note: In some years, gases can have negative consumption values. This occurs when countries destroy or export gases that were produced in previous years (i.e. stockpiles).

1. **Ozone-depleting tonnes (ODP tonnes):** Ozone-depleting tonnes measure the total potential of substances to deplete the ozone layer. Some substances that deplete the ozone layer are 'stronger' than others, meaning one tonne will cause greater damage than one tonne of another. ODP tonnes are calculated by multiplying a substance's emissions in tonnes, by its 'ozone-depleting potential.' Ozone-depleting potential measures how much depletion a substance causes relative to CFC-11, which has a value of 1.0. If one tonne of a gas caused twice the depletion of CFC-11, it would have a potential of 2.0.

FIGURE 4.2

How are we solving this problem? We are not relying on "the invisible hand" of the "free market" to magically fix things. There is concerted government regulation and industrial policy. Neither are we requiring people to relinquish material goods or make sacrifices. We are not shrinking the economy. Indeed, production and consumption of air conditioning and aerosols have grown enormously since the 1980s. Instead, we use science to identify problems and then we use innovation to develop alternative chemicals that perform the same intended functions without the unintended harms. That is modernity solving its own problems. It is green growth. We could tell the same story with other examples such as air pollution in London, a city that is much cleaner and wealthier now than it was in the mid-twentieth century.

The logic behind green growth goes like this:

a. In seeking the goal of modern development (well-being as material abundance, security, and convenience), people use their intelligence to create technologies. We turn the mere potential value of nature into useful things like CFCs.

b. Initially, we have only a fairly limited range of intended outcomes in mind. In this case, we want comfortable air and ways to refrigerate food to forestall spoilage.

c. Yet simultaneously, we use science to monitor the broader impacts of new technologies.

d. When science discovers an unintended negative impact, we use politics and ethics to steer the techno-economy in a more beneficial direction.

This process is the rocket ship taking off as it temporarily causes harm but then learns to continue its upward trajectory in ways that are less damaging. The eco-modernists don't think of natural resources like a finite number of gumballs in a bowl. Rather, they see the human mind as the ultimate resource, capable of generating more resources by using the raw material of nature in ever smarter ways. After all, people in this case want cold air— they don't want CFCs. So, you just find a way to produce cold air without the harmful chemicals. Similarly, you can provide streaming entertainment without copper cables. You can provide electricity without coal, etc.

    Economists will tell this story via the crucial concept of environmental externalities. An externality is an unaccounted-for impact of production and consumption. The ozone hole is an external cost that is not factored into the price of production and consumption. So, the "private" price of, say, an air conditioner is lower than its "social" cost. The prices of goods drive supply and demand, so if they do not reflect the true costs of products, we have a problem. People will produce and consume in unsustainable ways. If a sustainable product is more expensive than the alternative, people will tend to buy the cheaper option.

    Climate change is a story of externalities. The narrowly-intended functions of our energy and agricultural systems are served well by fossil fuels. Yet GHGs have been left out of the economic equation. This means that the costs of extreme weather events, other climate harms, and the health impacts of burning fossil fuels are not fully included in the prices consumers pay (see Chapter 11). The logic of green growth is to account for such externalities (unintended impacts) in ways that foster yet more growth. Proponents of green growth say that this is the only practical and ethically responsible way to proceed.

The Case for Green Growth

The eco-modern climate scholar Roger Pielke (2010) frames our situation in terms of an "iron law" of climate policy: "When policies on emissions reductions collide with policies focused on economic growth, economic growth will win out every time." He thinks this is a fact every bit as constraining as the second law of thermodynamics.[1] In other words, growth is an imperative, not a choice. He argues that we will "not reduce emissions by willingly getting poorer." Nearly everyone is striving for more well-being along the modern development model. Any leaders that try to limit or oppose that will fail. Pielke says, "Calls for asceticism and sacrifice are a non-starter."

That sets a strict boundary on what we can pragmatically do to solve the climate problem. If growth is a foregone conclusion, our only hope is green growth via the process described above. The goal is to quickly reach "peak impact," especially peak emissions of GHGs, and then breakthrough as the rocket reaches escape velocity. It doesn't need as much fuel or resources (environmental degradation) to propel it much further distances (development). It can then cruise along a path of growth that is increasingly sustainable because it will be less and less impactful on the environment. Economic growth and techno-scientific innovation are "enabling constraints" that give us several powers to tackle climate change but also box us into only certain ways of approaching the problem.

We can push the argument further to say that green growth is a *moral* imperative, not just a political or economic imperative. After all, most of the forecasted economic growth this century will occur in the developing world where many people are poor and lack access to basic goods such as clean water, healthcare, education, and reliable electricity. Eco-modernists argue that modern living standards are crucial to human dignity and the only way to achieve them is through more of the evolving and growing process of modern development. So, green growth is climate justice.

They further argue that hardly anyone in the developed world wants to give up their way of life. And even if a few individuals want to pursue a life of simplicity, this is not a practical political project for everyone. We are locked into a way of life with no limits. We didn't stop at horses; we developed cars. We didn't say "enough" at radio. We developed television

---

1   Very basic laws of physics governing heat, energy, and so on. (Details of the Second Law are complicated and don't matter here.)

The Green Growth Model

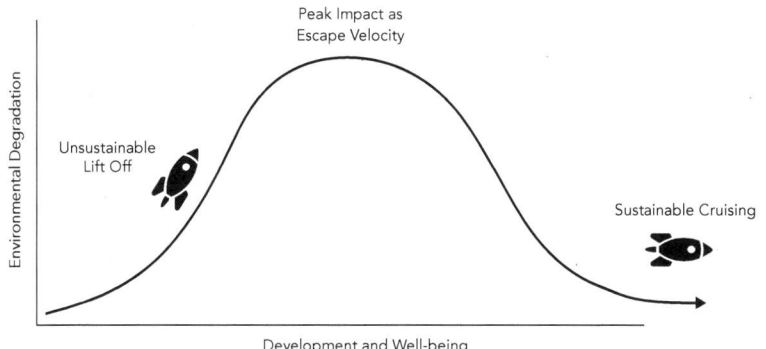

and the internet. We don't say "enough" to medicine just because we reach a certain average life-span. We keep going to 3D-printed organs, genetic engineering, and beyond. The same process—or imperative—governs climate change. We will push on to electric cars, vertical agriculture, lab-grown meat, precise fermentation, and space colonization.

This just is progress—the moral imperative to conquer nature for security and convenience. It is also worth pointing out that as affluence grows, the rate of per capita GDP growth declines. So, global GDP will likely eventually level off or grow only very slowly. And, recall that many affluent countries have already started to decouple economic growth from GHG emissions, so green growth is already proving possible.

To conclude, we are on a rocket and there is no way to turn around and land safely. With eight billion people on the planet, trying to go back to local forms of subsistence would be ecologically and economically devastating. We are dependent on this rocket that is our high-tech, high-energy, globalized civilization. We either accelerate and break through ... or we crash.

Degrowth

At the Youth4Climate Conference in 2021, climate activist Greta Thunberg delivered a scathing critique of green growth. She started her speech by seeming to applaud the green growth agenda of more jobs, green infrastructure, a smooth transition to a low carbon economy, etc., but then she changed tones, "There is no planet B. There is no planet Blah—blah, blah, blah, blah, blah, blah.... Green economy—blah, blah, blah" (see Kolbert 2022). In other words, it is all talk. All this talk of green growth and a grand energy transition and meanwhile GHG emissions just keep climbing and poor nations keep suffering.

Green growth is the way our major institutions like the IPCC frame the climate problem. Yet something interesting happened in the IPCC's sixth major assessment report, which was released in 2022. For the first time, they started talking about degrowth (see Parrique 2022). More generally, radical views like those of Thunberg are getting more serious consideration. In this section, we'll look at the logic of degrowth and the arguments in its favor.

The Logic of Degrowth

When the economy shrinks as a result of unplanned recession or depression, it causes lots of suffering: unemployment, homelessness, and hunger. Scholars who call for planned degrowth begin by noting that this is not what they have in mind. Consider a common definition of degrowth: "equitable downscaling of throughput [energy and resource flows through an economy], with a concomitant securing of well-being" (Kallis et al. 2018). The idea is to reduce our footprint by downsizing production and consumption in ways that are fair, safe, and desirable.

Crucially, "well-being" here indicates a vision of human flourishing that differs from the current predominant vision of growth. Recall that "development" means efforts to improve human well-being. So, degrowth is an attempt to redefine development and well-being in ways other than ever-growing gross domestic product (GDP). As the 2022 IPCC report notes, GDP can "be an inadequate metric for gauging well-being" (ch. 18, 19). It goes on to suggest that "prosperity" can mean "equity in well-being grounded in unanimity over shared goals and resources, rather than individualism, and economic and social and technological progress grounded in stewardship and care, rather than exploitation" (20).

To the "iron law" of climate policy, we might reply with the "tungsten law" of change (tungsten being the strongest metal that exists). All things change. The imperative of economic growth has only been true of one kind of human society and only very recently in human history. Most human cultures have found ways to flourish that are not dictated by the iron law. They have understood and related to the natural world in very different ways.

Yet we are seemingly locked into a commitment to growth premised on an understanding of nature as "resource." Maybe it is not an unbreakable law but it is very strong, like a tractor beam.[2] How could degrowth happen? In broad terms, it would require reimagining nature, sharing wealth, and simplifying life. Redistribution of existing wealth would be emphasized above greater production of more wealth. The IPCC report highlights several more specific policy mechanisms that could be used to accomplish degrowth objectives. These include:

a. A "cap and share"[3] framework for distributing GHG emissions permits on an annual declining basis designed to keep total emissions within the remaining global carbon budget.
b. Universal basic income and work-sharing to guarantee livable wages and full employment.
c. Shifting taxation burdens from income to resource and energy extraction.

In economics, the growth model of development and well-being has long ruled. But that is shifting now with a new generation of economists. They argue that sustainable development requires a sense of *sufficiency*. This would be a goal with metrics telling us not just when we have too little development but also when we have *enough* and even *too much*. The economist Kate Raworth (2017), for example, created a visual framework for sustainable development called "doughnut economics."

The hole in the doughnut is a zone of shortfall where development goals for human well-being are not met. The outer edge of the doughnut is defined by the ecological ceiling, indicating environmental thresholds that should limit growth. Beyond those thresholds is the zone of overshoot, where excessive impacts create dangers. One of these dangers is the climate. The sweet spot in the middle, the doughnut itself, is the "safe and just space for humanity," which is labeled the "regenerative and distributive economy." Note how this is grounded in metaphors of limits or boundaries like the footprint.

---

2   A science-fiction beam that can lock objects at a distance, or pull or push them (think *Star Trek* or *Star Wars*).

3   An economic proposal intended to bring ecological benefit. A "cap" places an equal upped limit on use of resources on all humans (which may be raised by purchase). All will "share" equally in benefits.

Doughnut Economics

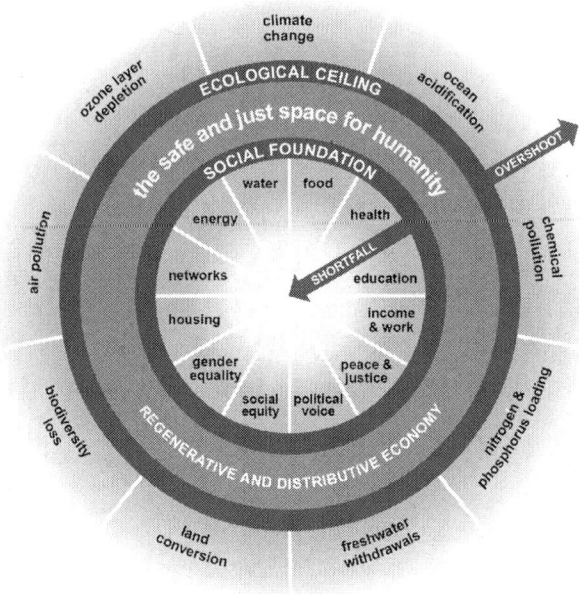

FIGURE 4.4

The Case for Degrowth

Let's survey five arguments against green growth and for degrowth. In each case, think about ways proponents of green growth might respond.

1. Technofabulism.[4] Green growth is just not realistic. Advocates for degrowth don't necessarily have to prove that green growth (global absolute decoupling) is impossible. They only need to argue that it cannot happen quickly enough to achieve the goals of climate safety like the 1.5°C target. Had more sincere efforts to decarbonize the economy started, say, in the 1980s, it may have been a feasible strategy. But that did not happen, so we are like the airplane on a short runway. It is time to hit the brakes.

---

4   Use of fantastical, unbelievable elements in one's story.

The European Environmental Bureau (Parrique et al. 2019) argues that there is "no empirical evidence" to support our faith in decoupling. It is just not happening on the scale and speed required. They conclude, that climate policy must now focus on "direct downscaling of economic production in many sectors and parallel reduction of consumption." The energy analyst Vaclav Smil (2019) has long made a similar argument that the grand energy transition to a decarbonized global economy is much more difficult than we imagine and will take much longer than we assume.

2. Green growth has already failed. We might push this argument further to say that the airplane is already starting to crash at the end of the runway. It's just that people in poorer parts of the world are the ones to feel the impacts first. Those in the developed world are temporarily shielded from the impacts. We are deluding ourselves in thinking that we are adapting to a warming world by, for example, building higher stilts for beachfront condos. This amounts to a passive form of climate denialism. What's really going on is that we are like the frog in the boiling pot[5] getting used to a situation that will soon be deadly.

3. Political naivete. Green growth is too susceptible to distortion by powerful economic groups. This is related to the above argument about how we are prone to comforting illusions. The narrative of "we just need more technoscientific innovation" is too easily captured and distorted by the carbon industrial complex (oil, coal, and natural gas companies). They will continue to conduct business-as-usual under a green disguise. As we'll see, arguably much of the activity around carbon credits is just such a smokescreen. They will exert their political power to ensure that favorable subsidies and other policies keep their products cheaper than the alternatives.

4. Injustices abound. Eco-modernists frame green growth as climate justice but that is wrong on three counts. First, their process approach

---

5 It's said that if you put a frog in cold water in a pot on the stove, and turn the burner on, the water will warm up slowly enough so as not to alarm the frog, who will, in consequence, sit there until it cooks. This turns out to be a myth.

to development assumes that people in the future will be wealthy and smart enough to solve the problems that we create. That justifies discounting the interests of future generations. But this doesn't seem to be how development is happening. In the US, for example, each new generation holds less wealth than previous ones did at their age. And younger generations are saddled with more debt. Second, and related, the development project tends to create the very poverty it purports to solve (see Escobar 1994). By destabilizing local ways of life and disembedding local economies, development forces people to become dependent on its geopolitical systems of finance, supply chains, and labor markets. Third, ecological limits mean that poorer nations can only have the resources for development if rich nations reduce their footprint.

5. Well-being does not grow indefinitely. Human happiness tends to level off after a certain threshold of material development and energy consumption (see Jackson et al. 2022). This makes sense—a billionaire is not millions of times happier than you just because they have millions of times more money. Numerous studies show that human development metrics plateau after a certain level of energy consumption (see Figure 4.5). So, why do we keep irrationally over-consuming when it doesn't make us happier, healthier, or freer and makes the climate more dangerous?

Maybe sufficiency via degrowth would increase well-being. After all, the development model of green growth tends to shape us into individualized consumers on one hand and faceless workers on the other. As development spreads, it tends to dissolve the coherent and engaging character of a world composed of focal things and practices. It disrupts traditions, communities, and cultures in ways that can be alienating. So-called development steals all the time we might devote to the serious pleasures of leisure (e.g., art) by exhausting us with work and distracting us with entertainment until it is time to work again. In liberating us from place, it tends to weaken the significance that special places might otherwise hold in our lives. It can supply necessities, but also create new needs, "false needs," in the process. At a certain point, economies can become over-productive, with products looking for markets and planned obsolescence forcing over-consumption. Is life really all about shopping, entertainment, and travel? If the availabil-

Nine Metrics for Human Well-Being and per Capita Energy Consumption

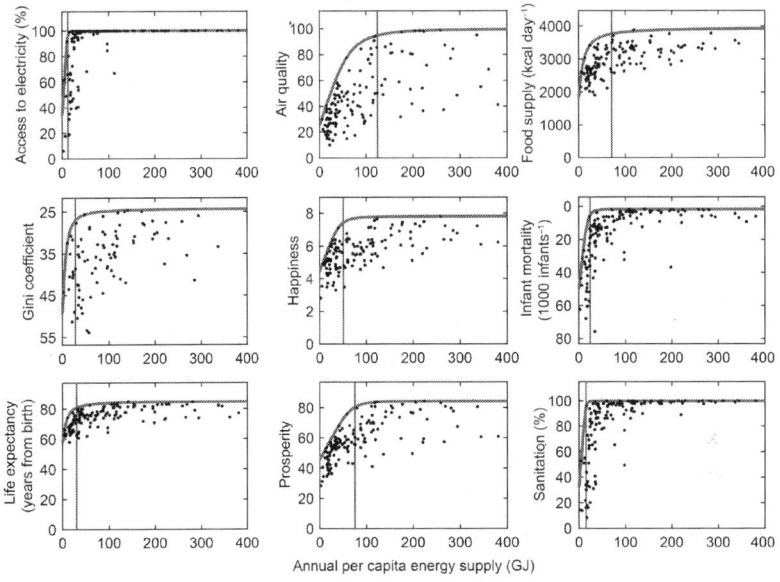

FIGURE 4.5

ity of pleasurable experiences is the goal, why not just plug us all into the "Matrix"?[6] And why are so many people in the developed world suffering from anxiety and depression?

Conclusion

This wraps up our section on "The Big Picture and Fundamentals." We are left with a disturbing ambiguity: Neither green growth nor degrowth seem particularly plausible. On one hand, green growth would require indefinitely expanding the development project on a finite planet that is already changing and degrading under the influence of our impacts. On the other hand, degrowth would require massive changes to structures and habits

---

6   In a science-fiction film with that name, a gigantic computer has all humanity trapped in its artificial reality.

that are deeply entrenched. I invite you to hold this tension in mind as we move forward.

Now it is time to turn to the climate sciences. From our big picture perspective, we know something about the place of the climate sciences—both their vital roles and their limitations. They help us to conceive of and measure the climate. They alert us to how we are unintentionally altering the climate. And although they cannot define the problem or dictate our choices, they can expand, evaluate, and clarify the alternative courses of action that we might take. Let's delve deeper into climate science: what do we know, how do we know it, and who is this knowing "we"?

Activities

1. Research the famous wager between the economist Julian Simon and the ecologist Paul Ehrlich. What was their bet and how does it relate to the green growth versus degrowth debate? Who won the bet and how did the winner and loser explain that outcome? The same contrast is in play between what the journalist Charles Mann (2018) calls "the wizard" and "the prophet." Who were they and what do these titles mean?

2. Consider the "doughnut economics" model. How might an eco-modern proponent of green growth respond to this? Could they say, for example, that we don't have to limit growth to stay under the ecological ceiling? After all, couldn't we just get more efficient with resource uses in ways that allow more wealth with impacts that fall within the safe zone of the doughnut?

3. Research the Green Revolution 2.0. How does this exemplify the logic and morality of green growth?

4. What ways can you re-envision progress, development, and well-being? What are the roles of spirituality or art in these imaginaries? How would science and technology fit in? Who would have what kinds of political power? What would the economy look like? How do we get from here to there?

References

Escobar, Arturo. 1994. *Encountering Development: The Making and Unmaking of the Third World*. Princeton, NJ: Princeton University Press.

IPCC. 2022. "Summary for Policymakers." In *Climate Change 2022: Impacts, Adaptation, and Vulnerability. Contribution of Working Group II to the Sixth Assessment Report of the Intergovernmental Panel on Climate Change*. Cambridge: Cambridge University Press.

Jackson, Robert, et al. 2022. "Human Well-Being and Per Capita Energy Use." *Ecosphere* 13, no. 4 (April): 1–10.

Kallis, Giorgos, et al. 2018. "Research on Degrowth." *Annual Review of Environmental Resources*, vol. 43, pp. 291–316.

Karlsson, Rasmus. 2015. "Three Metaphors for Sustainability in the Anthropocene." *The Anthropocene Review* 3 (1): 1–10.

Kolbert, Elizabeth. 2022. "Climate Change from A to Z." *New Yorker*, November 21. https://www.newyorker.com/magazine/2022/11/28/climate-change-from-a-to-z.

Mann, Charles. 2018. *The Wizard and the Prophet: Two Remarkable Scientists and Their Dueling Visions to Shape Tomorrow's World*. New York: Penguin Random House.

Meadows, Donella, Jorgen Randers, and Dennis L. Meadows. 2004. *Limits to Growth: The 30-Year Update*. Chelsea, VT: Chelsea Green.

Parrique, Timothee. 2022. "Degrowth in the IPCC AR6 WG II." https://timotheeparrique.com/degrowth-in-the-ipcc-ar6-wgii/.

Parrique, Timothee, et al. 2019. *Decoupling Debunked: Evidence and Arguments against Green Growth as a Sole Strategy for Sustainability*. Brussels: European Environmental Bureau.

Pielke, Roger. 2010. *The Climate Fix: What Scientists and Politicians Won't Tell You about Global Warming*. New York: Basic Books.

Raworth, Kate. 2017. *Doughnut Economics: Seven Ways to Think like a 21st-Century Economist*. New York: Random House.

Shellenberger, Michael, and Ted Nordhaus. 2011. "Evolve." *Orion*. https://orionmagazine.org/article/evolve/.

Smil, Vaclav. 2019. *Growth: From Microorganisms to Megacities*. Cambridge, MA: MIT Press.

PART 2

Climate Sciences

CHAPTER 5

Who Knows?
Institutions and Norms

Global Atmospheric CO₂ Concentrations

Atmospheric carbon dioxide (CO₂) concentration is measured in parts per million (ppm).

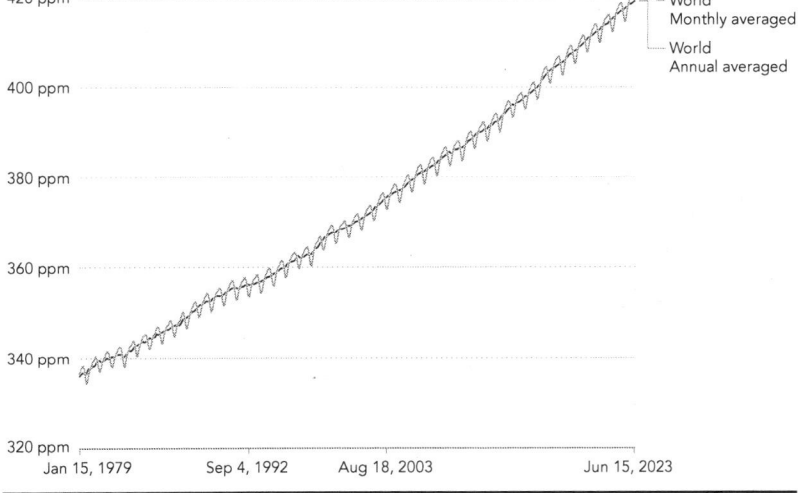

FIGURE 5.1

S cientific figures reveal lots of information at a glance. The Keeling Curve above shows atmospheric concentrations of CO_2 rising from about 340 ppm in 1979 to 420 ppm in 2023. This is one of the most important scientific records of the twentieth century, because it confirmed anthropogenic contributions to the greenhouse effect and brought public attention to global warming. The Keeling Curve captures the impacts of industrialized humanity in one line that is plainly intelligible. It makes climate change visible, disclosing an otherwise imperceptible reality.

Yet scientific figures also hide things from us. People make them, but we don't see that process. The line doesn't show the American scientist Charles Keeling struggling to secure funding for measurements at Mauna

Loa Observatory in Hawaii. We don't see the development of the infrared spectrophotometer. And hidden are debates about how to calibrate the instruments, account for sources of contamination, or compare data sets. A network of people and tools is hidden behind the image of that curve. We often think of technology as applied science, but science is also applied technology: we would not have knowledge without tools.

Hidden behind that line is also the historical drama of science where each generation pieces together more of the puzzle. Keeling was indebted to the work of the Swedish scientist Svante Arrhenius, who in 1896 used physical chemistry to argue that the changing proportion of CO_2 in the atmosphere would increase temperatures. In the 1850s, the American scientist Eunice Newton Foote was experimenting on the warming effect of sunlight on different gases. She was following clues left by the French physicist Joseph Fourier, who calculated that, given its distance from the sun, the Earth should be colder than it is. He was the first to think that gasses in the Earth's atmosphere could act like panes of glass to cause warming.

Fourier was building on the work of the English scientist Isaac Newton, who had (around 1700) described the mathematical laws operative in the natural world, including laws related to heat. And Newton was inspired by the French philosopher René Descartes who (in 1637) imagined a cosmos governed by universal laws and the scientific method for understanding the world so conceived.

Anthropogenic climate change has been one of the most comprehensively tested theories in science (see Winsberg 2018). This section of the book offers an overview of the climate sciences to help us work on our verification skills (evaluating the truth of statements). The three chapters are organized by a taxonomy of science that I find useful:

a. Science as institutions and norms (Chapter 5)
b. Science as bodies of knowledge and uncertainty (Chapter 6)
c. Science as methods and tools (Chapter 7)

This gives us a definition:

- The climate sciences are (a) norm-governed activities done by credentialed professionals working at institutions that are vested with social trust who (b) aim to extend validated knowledge (and characterize uncertainties) that explains and predicts the workings of the global climate system (c) using systematic studies that entail theory, observations, measurements, and models.

I emphasize *sciences* (plural) because of the diverse fields involved—climatology, paleoclimatology, physics, atmospheric chemistry, ecology, economics, sociology, and many more. There are many norms, institutions, bodies of knowledge, and methods involved. What holds them together as "science"? What demarcates science from non-science or pseudo-science? We'll first look at two answers: science is a worldview and science is our "epistemic constitution," which hinges on the norms governing science. Then, we'll turn to a survey of climate science institutions.

Science Is a Worldview

The climate sciences are rooted in the modern scientific worldview sketched by Descartes. This worldview is non-narrative, naturalistic, and mechanistic. That is to say that it explains phenomena by appeal only to physical nature and reason, not to super-natural actors, and it pictures reality as matter and energy governed by universal laws. The key distinction here is between science (or natural philosophy) and myth.

This is abstract, so let's illustrate it with the phenomenon we are most interested in. The "climate system" is composed of gasses, liquids, and solids that are atoms and molecules moving with interchangeable and quantifiable energies. Thus, for example, science pictures terawatts[1] of solar radiation striking the Earth's surface and not, say, a sun god driving a chariot across the sky. The climate system is governed by physical laws of force, gravity, pressure, magnetism, volume, heat, etc. Thanks to Arrhenius, Newton, Foote, and others, we know that the greenhouse effect is explained by the molecular configuration of certain gasses reacting to long-wave (roughly 10–30 μm) infrared energy.

In other words, the sciences define our concept of "the climate" and explain how our activities impact the climate system. In these ways, science sets the terms of debate for climate change, establishing the official explanatory framework or worldview.

---

1   A unit of energy = one trillion (10^{12}) watts.

There are other worldviews. For example, Indigenous Studies scholar Vanessa Watts (2013) notes that many North American Indigenous worldviews are different. She describes conceptions of Place-Thought, where each unique landscape possesses its own form of agency in contrast to the universal laws of modern science. Similarly, the climate scholar Daniel Wildcat (2009) argues that "indigenous realism" pictures reality as a complex web of life that cannot be reduced to resources, facts, and figures. This worldview foregrounds relationships with the land and responsibilities across species in ways that differ from the modern scientific worldview.

So, modern science is rather unusual in some respects—not many worldviews developed by cultures have pictured reality governed by universal, calculable, impersonal laws. It is far more common to find mythical worldviews, where the actions of deities of various sorts explain how things are. In other ways, though, the modern scientific worldview is just a precisioning of human intelligibility. Humans naturally understand things by seeing them within frameworks of lawful regularities. We notice the rhythmic migrations of bison, for example, or the bloom, grow, and fade pattern of the seasons and of each mortal life. We track the motions of stars and the sun. Such a view gives us useful knowledge, helping us predict and manipulate the world around us.

So, we subsume particular events under generalities, which is a move from diversity to sameness as each unique thing is treated the same as other instances of a type. Attention shifts from the individual bison to the herd. What makes modern science so powerful is how it pushes this logic to the extreme (Borgmann 1984). Bread and bronze are the same when analyzed on an atomic level. The falling apple, the cannonball, the Earth, the Sun, and the comet are all instances of the same kind: objects in motion governed by universal laws. The comet is no longer an omen sent by deities. Similarly, humans are products of the same evolutionary forces that have generated all species.

It is this way of looking at the world that has both (a) unleashed the technological powers driving development or the Anthropocene and (b) given us the tools to warn us about climate change and other unintended consequences of those powers.

As soon as Descartes imagined this worldview, he recognized that it was disorienting. It is like being lost in a forest with no trail. Everything is the same in all directions: matter in motion. To pick out significant markers and find our way, we need other ways of viewing the world. To see a giraffe on the savannah, for example, requires more than science to focus our attention on those units of analysis (giraffe and savannah) rather than, say, this or that cluster of molecules. Science can so powerfully *explain* the world, but it does not *articulate* it. It doesn't tell us what is meaningful or relevant. Science is a non-narrative form of rationality but we are narrative creatures. Our minds crave meaning and leap at it. When we look at the Keeling Curve, for example, we can't help but weave it into a story. Climate literacy entails understanding these entanglements of the scientific worldview and other ways of seeing the world—ways that are humanistic, religious, cultural, and political.

Science Is Our Constitution of Knowledge

Science is a quest for knowledge about "objective reality." The Keeling Curve exists because there really are more and more CO_2 molecules in the atmosphere. Objective reality indicates a mind-independent world, but our access to this world is profoundly influenced by the way our sense organs and brains work, and by our cultural upbringing. It's hard to even know what "direct" access would be: the "view from nowhere" or a "God's eye view"? The point is that we cannot step outside of ourselves to know things "as they really are." We have to get at reality with our minds, senses, and tools (which often act to extend our senses), and we all have our own perspectives (remember our mental models!). So, what we claim to know is always indirect or mediated in these ways. When we disagree, then, how can we decide who is right? This social function of deciding on reality is another good way to understand what science is.

The journalist Jonathan Rauch (2021) argues that objective reality is a set of propositions that have been "validated in some way." In many cultures, oracles and priests perform this validating role. They tell people, for example, how the Sun relates to the Earth or what kind of omen a comet portends. Natural philosophers (the precursors to modern scientists) lived on the margins of these other sources of power-truth. When they challenged authorities, the outcome was often grim—recall the fates of Socrates and Galileo.

The modern institutionalization of science moved it into the center of society, conferring authority on scientists (Ben-David 1984). As science became an institution, it codified norms for its self-regulation and quality control: credentialing, peer-review, and the replication of studies to either confirm or falsify findings. We also call this professionalization. It is how scientists distinguish themselves from non-scientists. Society grants science autonomy in exchange for producing useful and accurate knowledge that is independent of tradition, partisan faction, and revelation.

Society can be riven by disagreements about "objective reality." As we have seen, there is not always consensus on the relevant facts or how to interpret them. But there is broad consensus that the sciences are where we turn to resolve questions about reality. We also turn to other institutions like the justice system, intelligence agencies, and journalism. Together, these decentralized institutions guided by norms of truth-testing form what Rauch calls our "constitution of knowledge." They establish the justified true beliefs (knowledge) that, ideally, guide collective decisions.

The term "constitution" can be read in two ways:

1. Noun: A framework establishing the principles and procedures for decision-making.
2. Verb: The processes of *constituting* knowledge.

Take as an example this knowledge claim from the IPCC (2021): "The likely range of total human-caused global surface temperature increase from 1850–1900 to 2010–2019 is 0.8°C to 1.3°C, with a best estimate of 1.07°C" (5). To say such knowledge is "constituted" does not mean it is fabricated or "made-up" out of thin air. It indicates an "objective reality." The point is that this reality is neither self-evidently obvious (like "the cat is on the mat") nor true by definition (like "a bachelor is an unmarried man"). Getting at the reality of a warming Earth took a vast network of people using instruments in norm-governed ways. That is the *constitution* of knowledge.

Norms are like rules of the game that define its identity. In ice hockey, for example, you have to use your stick to advance the puck. If the players throw the puck back-and-forth with their hands, then they are not playing ice hockey. Likewise, if someone fabricates or falsifies data, plagiarizes others, or draws conclusions from wild speculations without evidence, then they are not doing science. Scientific norms define the legitimate methods for obtaining knowledge. They tell us what counts as science and how its results are validated, that is, confirmed or falsified. They set the standards for collecting and evaluating evidence and drawing inferences and conclusions from that evidence.

Philosophers and scientists have long debated just what the norms of science are (see Longino 1990). The sociologist Robert K. Merton (1942) argued that the goal of science is the "extension of certified knowledge." To accomplish this goal, four norms are necessary—to violate them means that one is no longer doing science. His norms go by the acronym CUDOS.

1. Communism:[2] Individuals cease to keep knowledge secret but systematically share it with other scientists and to some degree with the public.

2. Universalism: Knowledge claims are evaluated by universal, pre-established, impersonal criteria, without reference to the nationality, religion, class, race, or gender of the scientists making them. And scientists rise through the professional ranks on the basis solely of their merit rather than personal or political ties.

3. Disinterestedness: This is institutional self-policing that holds scientists accountable only to their peers (armed with scientific standards of truth-validating) rather than to their own wishful thinking or the interests of other parts of society. This accounts for the centrality of peer review.

4. Organized Skepticism: All knowledge claims are subject to rigorous community scrutiny. It requires the temporary suspension of judgment and detached examination of beliefs, even those deemed by some to be beyond the scope of rational analysis. This also implies that all knowledge claims are provisional, that is, subject to further testing. It means that even the assessment reports produced by the IPCC are not sacred texts. They are useful guides for action and snapshots of expert knowledge at a certain time, but they too must remain open to dissent and criticism.

The constitution of knowledge is like a funnel. The wide end represents political discourse broadly conceived, including social media. This aspect is vital for democracies. Their governments give broad latitude for free speech and idea pluralism. The institutions and norms of science (and the other truth-testing professions) act as the narrowing part of the funnel. Like gatekeepers, they only allow certain knowledge claims to pass. This too is

---

2 Merton is clearly not talking about the political system of the former Soviet Union. He just means community-oriented.

The Constitution of Knowledge

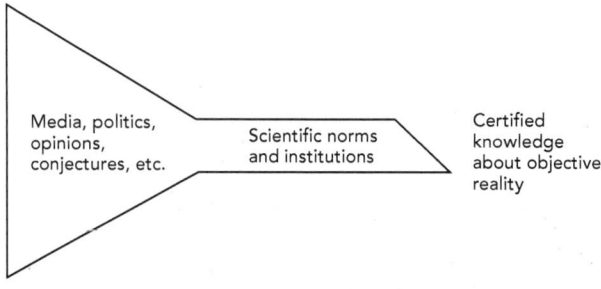

FIGURE 5.2

vital for democracies that need independent, evidence-based justifications for policy-making. This is why governments develop bureaucracies full of experts. Of course, as we've noted, there will still often be disagreements among experts, uncertainty, and room for different legitimate interpretations of data. This raises again the need to keep in mind the dangers of both false equivalence (not everyone is qualified to evaluate knowledge claims and not all knowledge claims have been equally tested by the constitution) and false binary (uncertainty and legitimately different interpretations and conclusions often exist—it's not always as simple as "science" versus "junk").

A social consensus on a norm-governed process for truth-testing carried out by trusted professionals is the best way we have for settling questions about reality. And this is why the climate sciences have grown and moved into the center of politics and policy as the urgency of climate change became more evident: society needs reliable knowledge, reasonable inferences, and sound characterizations of uncertainty (independent from various interest groups) to guide decisions.

A Brief Survey of the Climate Sciences

In 1988, the NASA climate scientist James Hansen testified before the US Senate. He said that scientists had been studying global warming for a long time. Until then, they had been cautious about attributing rising temperatures to human influence. But now he felt confident that the warming they were tracking was not a natural variation. Dr. Hansen said, "It is time to stop waffling...." His testimony was a defining moment: climate change was suddenly all over the media, garnering public concern and attention from politicians.

Hansen's testimony is a watershed that can divide this brief historical survey into two parts. Part one tells the pre-history of the institutions of climate science when knowledge was being discovered largely out of view of the public eye and under the assumption that humans were too insignificant to impact the Earth. The second part tells how the relevant institutions developed and moved toward the center of society as science helped us understand that we had become geologic agents of climate change.

Part One: The Discovery of a Changing Climate

A thousand years ago, the Chinese natural philosopher Shen Kuo (1031–95) developed a theory that is the cornerstone of any climate science: the Earth is extremely old and dynamic. Any given place was once very different, with different animals, plants, and weather patterns. Recall the Earth Calendar: humans have only been around for the last half-hour of the last day. Shen used observations and evidence to support his theory. He wrote, for example, of a marine fossil discovered far inland, and he noted the discovery of petrified bamboo deep underground in an ecosystem that was no longer habitable by bamboo.

Shen's inquiries illustrate the first of four ingredients that are key to the early history of the climate sciences.

1. *Wonder*: A courageous and creative openness to wonder is the beginning of climate science (like all science). People need to think beyond received wisdom and their day-to-day experiences. This happened in fits and starts with various individuals in different cultures. In the West, this wondering was encouraged and institutionalized during the eighteenth century as natural philosophers began piecing together clues of Earth's deep, evolving history. Much of this activity was religiously motivated, like attempts to find evidence of the Great Flood as told in the Bible. Many, like the French naturalist Comte de Buffon, though, challenged biblical accounts of the history of the Earth. He speculated that the Earth was 75,000 (not 6,000[3]) years

---

3   In 1650, the Irish archbishop James Ussher published his calculations, based on the Bible, of the day of Creation: October 22, 4004 BCE. Ussher's chronology was generally accepted until it came under serious scientific attack a hundred years later.

old. His ideas were condemned by Church authorities, and he was forced to publicly recant his work. This shows how the threat of persecution needs to be removed or at least severely dampened in order for science to operate as a guide at the center of society, rather than as a heterodox voice on the margins of society.

2. *Communication*: Science is a form of collective intelligence. No individual could ever piece together all aspects of the climate. Through the eighteenth century, communication largely occurred in the informal Republic of Letters, a vast network of long-distance intellectual exchanges covering a variety of topics. In the nineteenth century, peer-reviewed journals (a genre dating back to 1665 with the founding of the Royal Society of London) became the main medium of exchange. Today, well over 30,000 climate change research articles are published in peer-reviewed journals annually.

3. *Institutional support*: In the nineteenth century, two institutional homes for climate science took shape. First, many universities turned to research—the production and certification of new knowledge, not just the curation of existing knowledge. This also meant more autonomy for "scientists" (a term coined in 1834) and the reduction of persecution and removal of religious controls over knowledge claims. Second, governments began to form scientific agencies, like the US Geologic Survey, which conducted research on climate. By the 1950s, climate change science was a recognized field of diverse disciplines. In the 1970s, private industry became a third institutional home of climate research. Exxon, then the world's largest company, funded a thriving climate research group. Into the 1980s, the oil industry and academia were on the same page about the reality of global warming and its potentially catastrophic consequences.

4. *Improved tools*: Cold War research and development funding was good for climate science. NASA began studying the Earth's climate with its first satellite launch in 1960. By the 1970s, satellite probes were exploring Venus, which improved understanding of our own climate system. Computers, although rudimentary, were starting to generate useful models of the climate system and helpful projections of the future climate under various scenarios. Scientists were

developing new tools to study the role of the oceans in the carbon cycle and the signatures of past climates (paleoclimatology), giving us more precise understanding of the factors driving climate change. And more longitudinal data was being collected from the massive Global Mean Temperature data set. By the late 1980s, this was starting to signal a shift in average global temperatures.

That brings us to the watershed of 1988. A growing scientific community armed with more powerful tools and data sets was studying climate change at numerous institutions. Until that point, this work had been on the margins of public consciousness and off the political radar.

Part Two: The IPCC and Climate Science in the Heart of Society

Hansen's testimony changed all that. The public came to see that global warming is happening. I can remember those days. Like many people, I looked at the development around me—more cars, buildings, electricity towers, etc.—and I had a general sense that all that fossil fuel use must be heating and changing the planet. But I didn't know if that was true. Hansen's testimony and the spotlight it threw on climate science confirmed my hunch. That's how many people felt. We started paying attention and demanding that our elected representatives do the same. Political leaders, in turn, looked to scientists for advice.

This raised a key challenge for scientists. Their work was spread across a growing, global, decentralized network of institutions. It was published in a wide variety of peer-reviewed journals across many disciplines. How could they organize it all in a way that would be useful for decision-makers? How could they distill the main message while also honestly communicating uncertainties and noting where different expert communities disagree with one another? What does "the science" actually "say" and what does it mean for action—who should do what, when, and how? In other words, how do we get from the accumulated knowledge and uncertainties of a non-narrative rationality to a guiding narrative for society?

This is why the United Nations endorsed the creation of the Intergovernmental Panel on Climate Change in 1988. Based in Geneva, Switzerland, the IPCC does not conduct original scientific research. Rather, in what is probably the largest peer-review process in all of science, it reviews existing scientific literature to provide information on the state of knowledge about climate change: causes, impacts, risks, and possible responses. The IPCC is governed by its roughly 195 member states, which elect a bureau of scientists to serve for the duration of an assessment cycle. The bureau selects thousands of experts from around the world to prepare its "assessment reports," which are published every five to seven years.

The assessment reports contain three volumes: 1) The Physical Science Basis; 2) Impacts, Adaptation, and Vulnerability; and 3) Mitigation of Climate Change. These assessment reports also correspond to the three Working Groups of the IPCC, which are supported by Technical Support Units (TSUs). Here is an organization chart of the IPCC.

IPCC Organization Chart

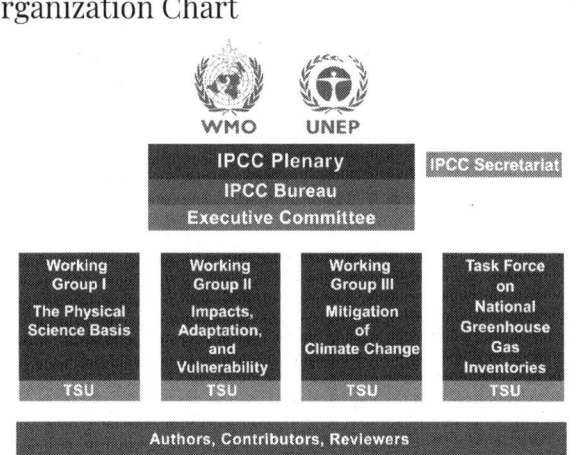

FIGURE 5.3

As you can see, the Task Force on National Greenhouse Gas Inventories (TFI), is the other major part of the IPCC's structure. The TFI develops methodologies and software for calculating and reporting national GHG emissions and removals. Having accurate and standard methods for reporting sources and sinks of GHGs is vital. Consider, for example, how policies might not actually achieve targets like the 1.5°C goal if we are under-counting emissions or over-estimating carbon removal.

The first assessment report in 1990 stated (in contrast to Hansen) that it was too soon to say that there was a definitive signal to attribute global warming to human causes. This changed with the second assessment report in 1995, which stated, "the balance of evidence suggests a discernible human influence on the global climate." By the sixth assessment report in 2021, the language was far more certain, "it is unequivocal that human influence has warmed the atmosphere, ocean, and land."

In 1992, the UN Framework Convention on Climate Change was created with the mission to "stabilize GHG concentrations in the atmosphere at a level that would prevent dangerous anthropogenic interference with the climate system." The UNFCCC is an international treaty that frames policy negotiations of the member nation states. It is the main mechanism by which IPCC reports gear into politics. I think of the IPCC and the UNFCCC as meeting together on a bridge between science and politics and policy.

IPCC and UNFCCC Meeting as Bridge between Science, Politics and Policy

FIGURE 5.4

Through the UNFCCC, nations use the IPCC reports as a foundation to build consensus and assign responsibilities for achieving its mission. Its supreme decision-making body is called the Conference of the Parties (COP), which meets annually to assess progress toward climate goals and to negotiate new adaptation, mitigation, and finance commitments. This process produces the nationally determined contributions (NDCs) for each party (e.g., nation), which are their non-binding plans for GHG emissions reductions.

The UNFCCC brokers agreements on overarching treaties. The 1997 Kyoto Protocol was the first implementation of measures under the UNFCCC. The Kyoto Protocol was superseded by the Paris Agreement (signed in 2015 and ratified in 2016). The main differences are that the Paris Agreement blurred what had been a strict divide between developed and developing nations and adopted the more ambitious 1.5°C target.

Generally speaking, any academic or governmental institution and peer-reviewed publication is a trustworthy source of information. The IPCC is not the only important or credible source of information about climate change. In the first instance, there are all of the institutions comprising the decentralized constitution of knowledge from which the IPCC draws for its assessment reports. In addition, there are national bodies like the US National Oceanic and Atmospheric Administration (NOAA) and international bodies like the International Energy Agency (IEA) that supply crucial scientific data and assessments. The World Climate Research Programme runs the Coupled Model Intercomparison Project (CMIP), which is a collaborative framework for improving climate models. And there are credible non-governmental organizations like Carbon Brief, that create useful reports and data visualizations. I also recommend the website Our World in Data for excellent data sets and data visualizations that are compiled from credible sources.

The number of agencies, institutes, websites, organizations, etc. can quickly get disorienting. And sometimes credible organizations disagree with one another on certain points. For example, the IEA and IPCC take a different approach to modeling possible climate futures. When in doubt about a source of information and analysis, I look at where their funding comes from and what biases that might indicate. I also look to see if my sources have been peer-reviewed, and I ask myself what sources well-respected scientists find useful and credible. When I see trusted experts relying on other sources of information, I consider that a pretty good litmus test.

Conclusion

As a worldview, modern science discloses reality in ways that create unprecedented explanatory powers and control over the natural world. As a set of decentralized, norm-governed institutions, science is at the heart of our constitution of knowledge. With their truth-testing standards, the sciences refine our sense of objective reality to guide decisions. The theory of climate change exemplifies this process as a wide set of claims about the climate system remains provisional yet also hones in on an increasingly certain and clear view of the big picture. The climate sciences have moved from the fringes into the center of society as this alarming picture has come into focus.

The modern development project has been outstripping our knowledge about its impacts. The IPCC seeks to rectify that imbalance. We no longer have the excuse of ignorance. Knowledge carries with it a moral imperative to act. Through the UNFCCC process, the global community has been pledging to respond to the science. And yet the global economy has not decarbonized at nearly the rate required to "avoid dangerous anthropogenic interference" and wealthy nations have not come close to fulfilling their pledges to help poorer nations adapt to a warming world. In the next chapter, we'll see that this is partly due to industry efforts to prolong a profitable (yet destructive) status quo. In later chapters, we'll see that failures to act quickly are also due to difficult moral and political questions that remain even once we set aside unwarranted denial and delay.

Activities and Questions

1. Read the article by Vanessa Watts and discuss whether and how modern science can be put into conversation with Indigenous worldviews. How can these other ways of knowing enrich our understanding and guide our action?

2. Research the norms guiding scientific conduct. What constitutes the "responsible conduct of research" or "good scientific practice"? How does the US Government define "scientific misconduct"? What do these norms for scientific integrity imply for the way scientists communicate uncertainty and urgency to the media?

3. The "Nongovernmental International Panel on Climate Change" (NIPCC) sounds a lot like the IPCC. But it is not. Look it up online and discuss. I don't think it is a credible source of information and analysis. In fact, I think it is very problematic. Why might I think that? Do you agree? Next, pick any climate science institution and ask yourself similar questions. How would you assess their credibility?

References

Ben-David, Joseph. 1984. *The Scientist's Role in Society: A Comparative Study.* Chicago: University of Chicago Press.

Borgmann, Albert. 1984. *Technology and the Character of Contemporary Life.* Chicago: University of Chicago Press.

IPCC. 2021. *Climate Change 2021: The Physical Science Basis. Contribution of Working Group I to the Sixth Assessment Report.* Cambridge: Cambridge University Press.

Longino, Helen. 1990. *Science as Social Knowledge: Values and Objectivity in Scientific Inquiry.* Princeton, NJ: Princeton University Press.

Merton, Robert. 1942. "Science and Technology in a Democratic Order." *Journal of Legal and Political Sociology* 1: 115–26.

Rauch, Jonathan. 2021. *The Constitution of Knowledge: A Defense of Truth.* Washington, DC: Brookings Institution Press.

Watts, Vanessa. 2013. "Indigenous Place-Thought and Agency amongst Humans and Non-Humans." *Decolonization: Indigeneity, Education, and Society* 2 (1): 20–34.

Wildcat, Daniel. 2009. *Red Alert! Saving the Planet with Indigenous Knowledge.* Golden, CO: Fulcrum.

Winsberg, Eric. 2018. *Philosophy and Climate Science.* Cambridge: Cambridge University Press.

What Do We Know?

Knowledge and Uncertainty

Global Cumulative Glacier Ice Loss

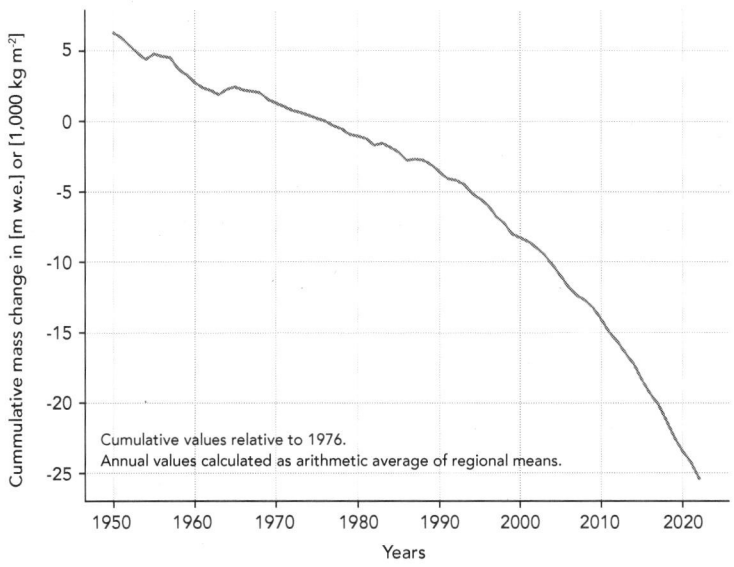

Cumulative values relative to 1976.
Annual values calculated as arithmetic average of regional means.

FIGURE 6.1

A round the world, mountain glaciers are retreating and disappearing. Those on Mt. Rainier in the US state of Washington provide examples. In 2023, a team of Mt. Ranier researchers used existing data sets, aerial images, models, orthomosaics (high-resolution images), and other techniques to calculate the changing glacial area and ice volume from 1896 to 2021 (Beason et al. 2023). They found that the glaciers are melting at an accelerating rate. The change in glacial area was -53.812 km² (-20.777 mi²), which is a reduction of 41.6 percent. The volume of glacial ice in snow in 2021 was 3.516 ± 1.231 km³ (0.844 ± 0.295 mi³), which represents a total reduction of 51.6 percent since 1896.

These are facts derived from the constitution of knowledge. Most people think of science as the collection of such facts. It is a body of knowledge— the stuff that is in textbooks, peer-reviewed journals, and scientific reports on phenomena like mountain glaciers. This body, so the common understanding goes, is composed of *facts* that have withstood the tests of scientific norms and institutions to become certified knowledge. Climate science textbooks, for example, present facts about glaciers, the carbon cycle, warming trends, the greenhouse effect, etc.

The growth of this body of science has been accelerating since the first peer-reviewed journal was founded by the Royal Society in 1665. Indeed, it is doubling in size now roughly every 17 years (Bornmann et al. 2021). (Is it troubling that glaciers keep melting as the science keeps growing?) Three characteristics of this "body" are important to note up front: it is alive, big, and nebulous.

1. Science is a *living* and dynamic body—growing new parts and sluffing off old parts. Some parts are more settled and some are more speculative or contested. Remember, people are *constituting* this body of knowledge. It is an ongoing process. It is also growing so fast that some skepticism is warranted (see Sarewitz 2016). Is it all really certified? How healthy is this body? This is a question about scientific integrity.

2. The body of science is a *giant*. You would have to read over 1,000 pages every day to keep up with the churn of peer-reviewed climate science. This means that all books, opinions, media stories, and even lawmakers' decisions are based on a partial grasp of "the science." No one, not even the IPCC, is a know-it-all. People necessarily make decisions under conditions of disagreement, partial knowledge, and uncertainty.

3. Since climate is a "hyperobject," the relevant body of knowledge is *ambiguous, diverse, and ill-defined*. We might need to research anything from the price of batteries to the production of solar panels to the intensity of heat waves to the economic impacts of a proposed policy and more.

The scope and ambiguity of the scientific body of knowledge introduces three fundamental questions. *What knowledge is relevant? When do we have enough knowledge? How should we contend with uncertainty?* We can add these to the key question from last chapter about how knowledge is constituted and what knowledge is credible.

Answers to these questions will vary as climate change appears in different contexts. Maybe you will need to know about the carbon cycle or maybe local zoning ordinances. So, rather than recite a laundry list of facts, let's work on skills to organize thinking as you consider the key questions: what do I need to know, when do I know enough, and how should I understand uncertainties? We can use the problem orientation framework—goals, trends, conditioning factors, projections, and alternatives—to guide us.

When we introduced this framework in Chapter 3, we focused mostly on goals. So, in this chapter and the next, I will emphasize trends, conditioning factors, and projections about the climate system as they relate to these goals. Keep in mind that the goals help determine what counts as relevant knowledge. Our emphasis here will be on the physical sciences. The last section of the book will bring the social sciences like economics and the humanities into the conversation. That will allow us to focus on alternatives as we squarely face politics, ethics, and policy.

Relevant Knowledge

Our main goals are *safety* (avoiding dangerous interference in the climate system) and *well-being* (flourishing, happiness, or development). So, what bodies of knowledge about trends, conditioning factors, and projections matter for these goals? This will depend on how you interpret those goals: what do they really mean? And it will depend on your context and scale. As a case study, let's take the UN's Sustainable Development Goal of Zero Hunger as an instance of human flourishing and think about, say, wheat.

Picture a simplified chain, where each preceding trend is a conditioning factor for the trend that follows.

Development → GHG emissions → Temperature → Wheat production → Prices → Hunger

As problem-oriented thinking, climate literacy asks: where is the problem? In other words, what is relevant knowledge? We know that trends for GHG emissions and temperatures are up. But is that a problem if wheat production is also up? Maybe we should look at the share of people globally who are undernourished. Isn't that ultimately what we care about? If so, Figure 6.2 suggests there is no problem here.

Prevalence of Undernourishment in Developing Countries, 1970–2015

The share of individuals that have a daily food intake that is insufficient to provide the amount of dietary energy required to maintain a normal, active, and healthy life.

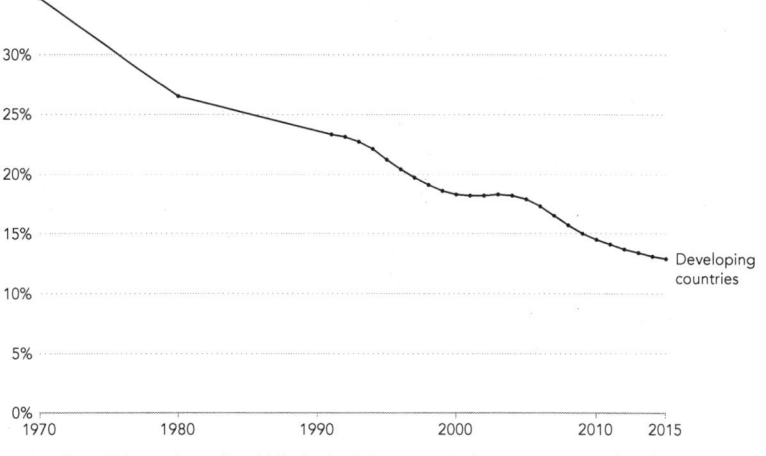

Note: Data from 1990 onwards is well-established within FAO estimates. Earlier estimates are significantly more uncertain.

FIGURE 6.2

Yet maybe projections suggest that hunger will increase if we don't rapidly and massively curtail GHG emissions. Recall, a problem means that our goals do not align either with existing trends and current conditions *or* with future projections. Then again, hunger often increases due to war or corruption that may have little to do with GHG emissions or temperature trends. But (*then again!*) climate change often does play a part in situations of political unrest, war, drought, and migration. What are the relevant conditioning factors here? How does climate change *appear* in this case?

The linear arrows above are, in reality, an enormously complex system of natural-technical-social trends and conditioning factors. It is a tangle—a web with feedback loops connected to things we have not yet considered.

Not pictured in the chain, for example, are genomic technologies[1] that can help wheat crops to grow in hotter and dryer conditions. Is a future that is hotter but better-nourished our desirable outcome? Of course, trends in democratic backsliding and fraying international relations might destabilize the innovation systems that such technologies rely on.

Again, the human mind can't help but try to make sense of all this by coming up with a simplified story—that mental model—and a problem-framing that will be based on selected facts. Our goal is to pay attention to how those models are made and how problems are defined, especially when it comes to the use of science. When people define the problem, what do they include or highlight and what do they omit or downplay? Climate politics cannot be just the sum total of climate facts. The report on Mt. Rainier's glaciers, for example, is 166 pages long. That's just some of the facts about just one mountain in a vast climate system. Which facts *matter*?

The point is that you don't just need knowledge (as if climate literacy was a matter of memorizing, say, all the facts about glaciers). More importantly, you need the meta-knowledge skills of seeing how knowledge is used. How are facts being marshalled to build arguments, make stories, and frame problems? (The word "fact" like "factory" has etymological roots meaning "to set or put" as one sets a building block in place.) This is the key to understanding our "post-truth" age where what matters is not so much the truth but *how* the truth is determined or constituted (see Fuller 2020).

The enormous body of scientific knowledge creates what the science policy scholar Daniel Sarewitz (2000) calls an "excess of objectivity." Many sciences produce many pictures of the world. It puts us in a situation like the parable of several blindfolded people feeling the body of an elephant. One has a hold of the trunk and says it is a snake. The other feels the side and says it is a wall. Another feels a leg and says it is a tree trunk. Yet another grabs the ear and says it is a fan. As noted, the IPCC was formed in part to deal with this issue: with so much complexity and so many perspectives (e.g., different scientific disciplines) how do we distill a common view, a shared story, and a collective problem-framing about climate change?

The metaphor of the elephant brings to mind our metaphysical point about how problems like climate change are framed: just what *is* it?! This is all part of the constitution of knowledge. Those who understand how to

---

1   Genomic technologies analyze and manipulate gene structures—the basis of inherited development. It is hoped that this can make food plants and animals feed us safely and more economically.

frame reality get to set the social agenda. If you don't have your eyes wide open to this dynamic, it doesn't matter how much data or information you have. The weightiest of scientific reports can land without making a noise; meanwhile, the smallest social media post can cause a political storm. You will lose the post-truth game if you keep insisting that we just listen to the facts. Which facts? How do you make them come alive with force? This speaks also to the challenge many scientists face now in trying to communicate about a growing climate "crisis" or "emergency."

Sensitivity, Detection, and Attribution

Developing the skills for assessing relevant and sufficient knowledge is one key to science literacy. Another important aspect is mastering some central concepts that can help you decipher climate science. Here, we'll focus on three of the most essential. We already know the key trends and conditioning factors, which are summarized in Figure 6.3 from the IPCC's sixth assessment report. It is "unequivocal" that increasing atmospheric GHG concentrations from human activities are driving up average global temperatures. And although "dangerous" levels are hard to define, we keeping inching into ever greater danger zones.

Now we can refine this basic understanding with some central terms in climate science. The first is *climate sensitivity*: "the warming we can expect when the concentration of carbon dioxide in the atmosphere reaches double what it was in preindustrial times" (Carbon Brief 2014). That roughly means 560 ppm, which is likely to happen around 2050. This is an important concept, because it is how the trend of GHG emissions functions as a conditioning factor for rising temperatures. It tells us how sensitive the Earth is to the radiative forcing (the heating) caused by our GHG emissions. If the climate system is hyper-sensitive (a high number), then the danger zone comes more rapidly. If the system is less sensitive, then we have more time.

As with most measurements in climate science, climate sensitivity comes in the form of a range that tells us where we think the actual number will be (this is why we will turn to uncertainty next). In its sixth assessment report, the IPCC put this "likely" range with "high confidence" between 2.5 and 4°C. Yet that doesn't mean the Earth will instantly be that much hotter once we hit 560 ppm. Much of the warming effects take a while to cycle from ocean to atmosphere, because the oceans actually absorb most of the heat

History of Global Temperature Change and Causes of Recent Warming

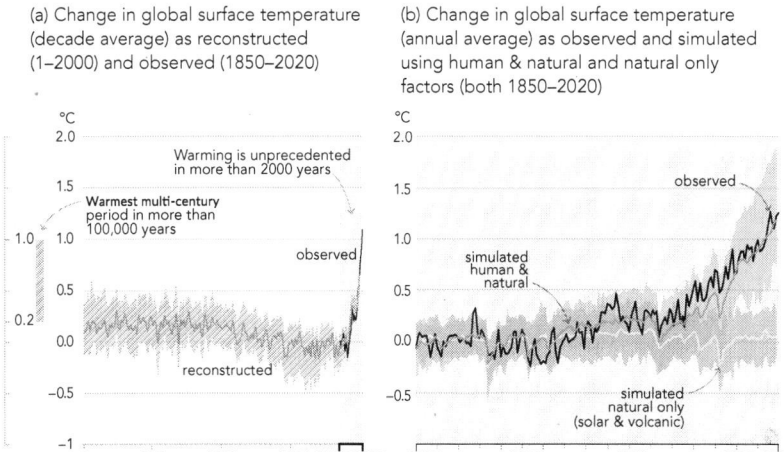

(a) Change in global surface temperature (decade average) as reconstructed (1–2000) and observed (1850–2020)

(b) Change in global surface temperature (annual average) as observed and simulated using human & natural and natural only factors (both 1850–2020)

Figure SPM. 1 | History of global temperature change and causes of recent warming

Panel (a) Changes in global surface temperature reconstructed from paleoclimate archives (solid gray line, years 1–2000) **and from direct observations** (solid black line, 1850–2020), both relative to 1850–1900 and decadally averaged. The vertical bar on the left shows the estimated temperature (*very likely* range) during the warmest multi-century period in at least the last 100,000 years, which occurred around 6500 years ago during the current interglacial period (Holocene). The Last Interglacial, around 125,000 years ago, is the next most recent candidate for a period of higher temperature. These past warm periods were caused by slow (multi-millennial) orbital variations. The gray shading with white diagonal lines shows the *very likely* ranges for the temperature reconstructions.

Panel (b) Changes in global surface temperature over the past 170 years relative to 1850–1900 and annually averaged, compared to Coupled Model Intercomparison Project Phase 6 (CMIP6) climate model simulations of the temperature response to both human and natural drivers and to only natural drivers (solar and volcanic activity). Solid lines show the multi-model average, and shaded areas show the *very likely* range of simulations.

FIGURE 6.3

in the near term from our GHG emissions. It's worth remembering that the climate is an open-ended, complex system involving massive thermal transfers of energy over different time periods.

The next key terms are *detection*, which means that we have spotted a trend, and *attribution*, which means that we are assigning a cause or conditioning factor (or "driver") to that trend. Look at Figure 6.3 again. On the left side, we see detection. Using data both from reconstructions and observations, scientists detect a trend of warming that is statistically significant. The right side of this figure shows attribution. Here, the IPCC is drawing from climate model simulations of the temperature response to both human and natural conditioning factors vs. the temperature response to only natural conditioning factors (fluctuations in solar radiation and volcanic activity). These results about trends and conditioning factors drawn from reconstructions, observations, and models continue to be confirmed and strengthened over the years as instruments improve and data accumulates. This is a big reason why the IPCC is now so unequivocal about attributing global warming to human industrial activities.

With the concepts of detection and attribution in your toolbox, you are well-positioned to understand scientific reports and media stories about climate change. These are often not just about the global trends in temperature. Indeed, "many changes in the climate system become larger in direct relation to increasing global warming" (IPCC 2021, 15). These include heat-waves, heavy precipitation, droughts, and reductions in sea ice, snow cover, and permafrost. In turn, these changes will impact more direct measures of our goals of safety and well-being, such as, extreme weather damages and shortages of clean water. Scientists are studying a wide range of natural phenomena to (a) *detect* how they are changing and (b) *attribute* the extent to which anthropogenic GHG emissions contribute to the changes they detect.

These detection and attribution sciences are getting better at seeing how human influences are tangled up with "natural" events (thus, such studies are often called "fingerprint assessments"). It is important, though, to read media stories with a critical eye and check them against actual scientific findings. Oftentimes the media is prone to exaggerate things in an effort to earn attention and clicks in a world saturated with information. So, be critical of headlines about climate change straightforwardly "causing" a specific event like a wildfire. Given the tangled systems involved, it is probably more accurate to consider it one conditioning factor among others, which then raises questions about how relevant different kinds of climate science and policy are for the context at hand.

Event attribution is an emerging area of science that measures how climate change affects specific extreme weather events—like a particular heatwave or tropical cyclone. The goal is to understand how much climate change affects the magnitude, intensity, and probability of any given event (see NAS 2016). Here, we can see how our discussion of causality in the scientific terms of "attribution" raises the moral, legal, and political term for causality, namely "responsibility." If we can know, for example, that a given heatwave was made X percent more damaging by anthropogenic climate change, and if we know that a given company is responsible for Y percent of global GHG emissions, then we could calculate the percent of costs that that company should pay. This kind of calculation may become more prevalent in the future. Like so much of the science-ethics-law-politics tangle of climate change, it will be steeped in uncertainty.

Characterizing Uncertainty

Characterizing uncertainty is a key aspect of science. So, what is uncertainty and how is it used—and abused—when it comes to climate science? Let's consider first two kinds of uncertainty that are not necessarily mutually exclusive:

a. Epistemic: incomplete knowledge
b. Psychological: doubt or a perceived lack of knowledge

The former is about missing knowledge concerning a situation, and the latter is about one's subjective degree of belief or confidence in belief. The "reasonable person" standard might come into play between these concepts. Just because you believed your lying friend that it was permissible to steal a box of cookies from the store doesn't mean you were justified in holding and acting on that belief. As the phrase goes: you should have known better.

Indeed, uncertainty implies a duty to know when it comes to climate change. Development has caused massive changes that would seem to both morally obligate us to anticipate outcomes (because of the increased stakes of action) *and* make such foreknowledge impossible (because we are entangled with enormously complex systems across vast stretches of space and time). We are unleashing powers that cause chains of events that we ought to foresee, but cannot.

Here is another useful taxonomy inspired by comments made by US Secretary of Defense Donald Rumsfeld in 2012:

a. Known knowns: things we know that we know.
b. Known unknowns: things that we know that we do not know.
c. Unknown unknowns: things we don't even know that we don't know.
d. Unknown knowns: things we (think that we) know but we don't really know.

The first two categories are fairly straightforward. An example of a known known is that human activities are warming the planet's average temperature. Climate tipping points are examples of important known unknowns in climate science. The IPPC defines these as "critical threshold[s] beyond which a system reorganizes, often abruptly and/or irreversibly" (IPCC 2021, 28). Such critical thresholds might be breached in the cases of thawing permafrost (which releases methane), melting glaciers, dying coral reefs, the collapse of the Atlantic meridional overturning circulation (AMOC),[2] and more. In all such cases, we know there are lots of unknowns about the dynamics involved.

The third category is a little trickier. Think about unknown unknowns: What might we not even realize about the climate or about human behavior that we don't know? Are there things we are not even looking at? They would be like blind spots in our field of vision—gaps that we are not aware of. Visual blind spots happen when the human brain behaves in such a way that we do not see *that we do not see*. Might we be operating a global civilization confidently unaware of something crucial?

The category of unknown knowns is the hardest to grasp. These are things we supposedly "know" (we have psychological certainty) but maybe they are actually (epistemically) wrong. In fact, our mental models might be these unknown knowns—our filters and assumptions that may be distorted or wrong. So, unknown knowns might be summarized as "comforting beliefs" that we don't realize (or don't want to realize!) are false.

This means that the uncertainty of climate change is itself uncertain. We don't know all the things we don't know about it and we can't really be sure that we actually know the things we "know." So, it is not like there is a finite gap in our knowledge that we will eventually fill-in such that we have a complete picture of the climate system, including our socio-technical systems.

In addition to distinguishing *kinds* of uncertainties, we can also identify at least four *sources* of uncertainty:

a. The inherent characteristics of the processes and phenomena being studied
b. Our incomplete or imperfect understanding of the things we study
c. The methodologies that we use
d. The contexts within which we conduct our studies (including the values and interests that shape judgments about data)

---

2   The AMOC is a system of ocean currents within the Atlantic Ocean that brings cold water south and warm water north.

In closed systems, epistemic uncertainties can be characterized with complete accuracy. Think, for example, of a game of dice or cards that displays random behavior yet outcome probabilities can be determined statistically. The climate system—like social and most natural systems—is not like this. It is open, meaning, it is not governed by laws that can be fully elucidated. And the climate displays chaotic behavior where small differences in initial conditions can produce widely divergent outcomes. This is why epistemic uncertainty is unavoidable: we will always have incomplete knowledge about cause–effect relations and we will never have certain knowledge about the future. Any "objective" or quantitative probabilities we state about the climate, then, are the result of "subjective" or qualitative judgments—albeit they may be reasonable expert judgments derived from lots of evidence and sound reasoning.

Epistemic uncertainties in open systems, then, can be expressed in qualitative and quantitative ways. Insurance companies use quantitative methods for setting premiums and meteorologists use them to produce weather forecasts like, say, there is a 50 percent chance of rain tomorrow. These methods often rely on the assumption that the past accurately foretells future behavior. Of course, we know that climate change can make that assumption false. This brings us to the terminology used by the IPCC.

Uncertainty in the IPCC and Climate Science Politics

The IPCC has developed a framework of calibrated uncertainty language to evaluate and communicate their findings (see Mastrandrea et al. 2011). It is crucial to understanding their reports. Take for example this text:

> A warmer climate will intensify very wet and very dry weather and climate events and seasons, with implications for flooding or drought (*high confidence*), but the location and frequency of these events depend on projected changes in regional atmospheric circulation, including monsoons and mid-latitude storm tracks. It is *very likely* that rainfall variability related to the El Niño–Southern Oscillation is projected to be amplified by the second half of the 21st century in the SSP2-4.5, SSP3-7.0 and SSP5-8.5 scenarios. (IPCC 2021, 19, emphasis added)

Here is another quick example from the same report: In the period 2011–2020, "Late summer Arctic sea ice area was smaller than at any time in at least the past 1000 years (*medium confidence*)" (8, emphasis added). I've highlighted the key terms, which are *confidence* and *likelihood*. The IPCC has a process for deciding on this language:

Process for Evaluating and Communicating the Degree of Certainty in Key Findings

This schematic illustrates the process for evaluating and communicating the degree of certainty in key findings that is outlined in the *Guidance Note for Lead Authors of the IPCC Fifth Assessment Report on Consistent Treatment of Uncertainties* (Mastrandrea et al. 2011)

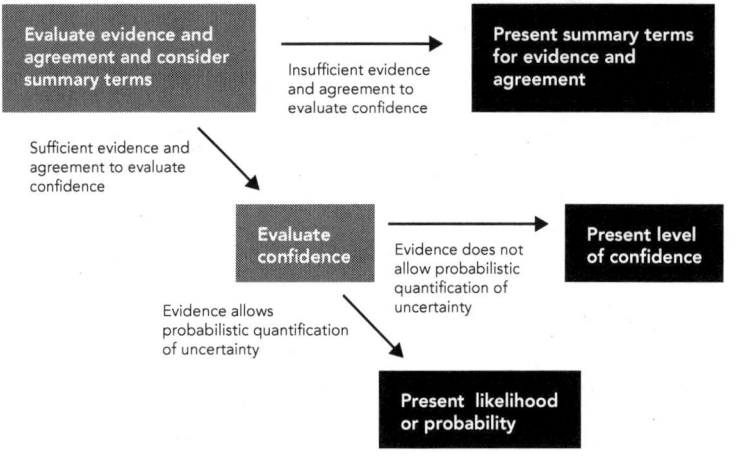

FIGURE 6.4

The first step is to evaluate the evidence and agreement for any given claim (e.g., about arctic sea ice extent or projected rainfall changes). The "summary terms" used for the type, amount, quality, and consistency of evidence are: limited, medium, or robust. For expert agreement, the terms are: low, medium, or high. The IPCC will limit itself to using these terms when there is insufficient evidence and agreement to evaluate confidence. The term "confidence" is a qualitative synthesis of evidence and agreement made by the relevant group of experts. This yields the "confidence scale" represented in figure 6.5.

The next step is to consider whether the uncertainties can be quantified probabilistically. If so, the experts characterize a claim in terms of degrees of likelihood ranging from "virtually certain" (99–100 percent probability)

A Depiction of Evidence and Agreement Statements and Their Relationship to Confidence

The nine possible combinations of summary terms for evidence and agreement are shown, along with their flexible relationship to confidence. In most cases, evidence is most robust when there are multiple, consistent independent lines of high-quality evidence. Confidence generally increases towards the top-right corner as suggested by the increasing strength of shading.

Agreement →

High agreement Limited evidence	High agreement Medium evidence	High agreement Robust evidence
Medium agreement Limited evidence	Medium agreement Medium evidence	Medium agreement Robust evidence
Low agreement Limited evidence	Low agreement Medium evidence	Low agreement Robust evidence

Evidence (type, amount, quality, consistency) →

Confidence Scale

FIGURE 6.5

through "about as likely as not" (33–66 percent probability) to "exceptionally unlikely" (0–1 percent probability).

Such systematic characterizations of uncertainty drawn from expert judgments are helpful for understanding the state of the science and for informing policymaking. Yet, they also clearly do not tell us what to do. Consider, for example, the issue of "tail risks," which are the chances of a loss due to a rare event. If an outcome is "very unlikely" but would be catastrophic, how much effort should we put into preventing it? In other words, we need to think not just about how uncertainty is characterized but also what it means for making decisions. We'll return to this with a discussion of risk in the next section of the book.

Profiting from Uncertainty

In his 1877 essay "The Ethics of Belief," the English mathematician William Clifford conjures a thought experiment. He imagines that the owner of a transatlantic ship is in a hurry to start the next voyage. Yet he is plagued by doubts—he knows his ship is old and not very well built. It's due for inspection and repairs. However, he talks himself out of these doubts. After all, the ship had sailed safely before and he "can put his trust in Providence" that all the people on board will have a safe voyage. The ship sinks, however, and everyone is killed.

Clifford draws a strong moral from the story: "it is wrong always, everywhere, and for anyone, to believe anything upon insufficient evidence." The shipowner "had no right to believe" on the basis of the evidence before him that the passengers would be safe.

There is truth to this. But it is not a simple matter. How much and what kind of evidence is sufficient? That's the key. The scientific knowledge we gather will never tell us when we have *enough* knowledge. Clifford's moral could be a recipe for paralysis, because we cannot escape uncertainty and, thus, we could always claim that we have insufficient knowledge. This kind of paralysis is itself a decision—a commitment to prolong the status quo. There is no option of not doing anything (as if we could stop time itself!). Delay in the quest for more knowledge carries its own risks as opportunity costs—the costs entailed in continuing with the status quo and the benefits we don't get had we chosen otherwise.

Since Clifford's time, science and technology have had success in making predictions and controlling the natural world. As a result, our expectations for certainty have risen. Politicians look to science to not just characterize but also *reduce* uncertainty. Yet when it comes to climate change although some key findings are now very well established, there is also an "uncertainty cascade" (Schneider and Kuntz-Duriseti 2002). The more we study the climate system, the more we realize how much we do not know. More knowledge often means more uncertainty.

And it means more fodder for debate and delay. Decisions about climate change are forward-looking and will create winners and losers (though how this all shakes out is also uncertain!). These decisions are also often couched in scientific terms. Thus, competing factions arm themselves with different studies and interpret knowledge claims and uncertainties in ways that frame the problem to suit their interests. Scientists tend to avoid wading into political debates for fear of appearing biased and hoping that eventually they will gather *enough* evidence to compel agreement and action.

Recall the "excess of objectivity" noted above by Sarewitz (2000). He uses this to portray an "iron triangle" that tends to keep climate science-politics spinning its wheels but not going anywhere. Politicians fund scientists to produce facts. Scientists produce an abundance of knowledge and uncertainties. Advocate groups can pick and choose their facts from this wealth to frame problems differently (different answers to the "what is it?" question). Or, in other words, the constitution of knowledge often leaves space for disagreement. Politicians often respond by asking for yet more science in hopes of eliminating this space by reducing uncertainty, and the cycle continues.

Iron Triangle

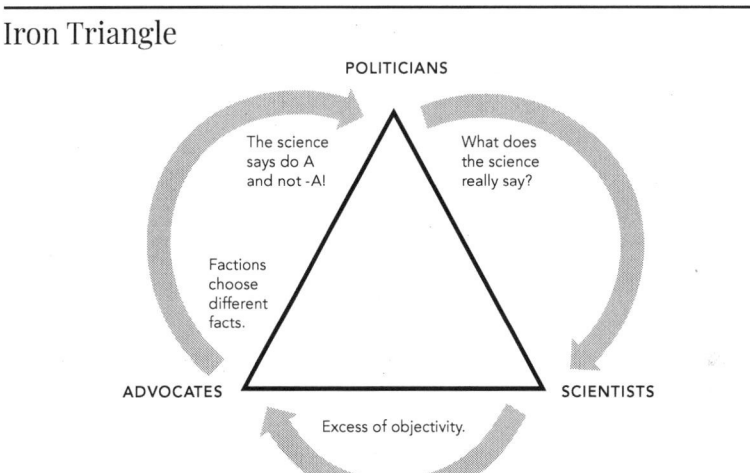

FIGURE 6.6

Yet, the excess of objectivity is not the only thing going on. There is also sheer power and big money. Around the time of James Hansen's testimony in 1988, the fossil fuel industry realized that action on climate change threatens their business. They adopted a strategy of emphasizing uncertainty in order to prevent action and, thus, prolong status quo dependence on their products (see Oreskes and Conway 2010). The fossil fuel industry and other major corporations formed the Global Climate Coalition (GCC), an international lobbying group dedicated to opposing climate action and challenging the science of climate change. The GCC produced reports that attacked the IPCC and sowed doubt about climate change. They also successfully lobbied against climate legislation such as the proposed BTU (British Thermal Unit) tax in the US in the early 1990s.

The key to this strategy is to emphasize the scientific norm of organized skepticism. Have we really done *sufficient* testing? Do we know *enough*? Don't we want to be absolutely sure before we act? If knowledge claims are always provisional, how can we be sure that any given controversy is really settled? In some contexts, these are fair and reasonable questions. But the fossil fuel industry has pushed them so far as to become incredibly reckless and immoral.

Conclusion

This chapter offered some tools for thinking critically about the enormous, growing, and ill-defined body of climate science. Rather than focus on specific knowledge claims (the last chapter showed us where to turn for those), our emphasis was on the skills for thinking through relevant, sufficient, and uncertain knowledge. We also covered some key concepts—sensitivity, detection, and attribution—that will help us understand, interpret, and evaluate climate science. This helps us to understand the uses and abuses of science in the politics of our post-truth age. Climate science literacy is not just filling your head with knowledge claims. It is also seeing how those claims are constructed and mobilized on shifting territories of contested authority.

Activities and Questions

1. Research the latest IPCC report and make a note of detection and attribution with confidence levels for all the phenomena they track (temperature, wildfires, drought, cyclones, ice, etc.). How does this compare to your expectations and to media coverage?

2. Research the "climate gate scandal." What were the accusations and did the scientists involved do anything wrong?

3. Watch the PBS *Frontline* Documentary "The Power of Big Oil." What does this history tell us about the place of science in a democratic, capitalist society? What lessons can we learn about the politicization of science (science treated as just another special interest) and the scientization of politics (values decisions disguised or treated as technical matters)?

References

Beason, Scott, et al. 2023. "Changes in Glacier Extents and Estimated Changes in Glacial Volume at Mount Rainier National Park, Washington, USA from 1896 to 2021." Ashford, WA: National Park Service. https://irma.nps.gov/DataStore/ DownloadFile/580543.

Bornmann, Lutz, et al. 2021. "Growth Rates of Modern Science: A Latent Piecewise Growth Curve Approach to Model Publication Numbers from Established and New Literature Databases." *Humanities and Social Sciences Communications* 8 (224): 1–15.

Carbon Brief. 2014. "Your Questions on Climate Sensitivity Answered." September 26. https://www.carbonbrief.org/your-questions-on-climate-sensitivity-answered/.

Clifford, W.K. (1877) 1999. *The Ethics of Belief and Other Essays.* Amherst, NH: Prometheus.

Fuller, Steve. 2020. *A Player's Guide to the Post-Truth Condition: The Name of the Game.* London: Anthem Press.

IPCC. 2021. "Summary for Policymakers." In *Climate Change 2021: The Physical Science Basis. Contribution of Working Group I to the Sixth Assessment Report of the Intergovernmental Panel on Climate Change.* Cambridge: Cambridge University Press.

Mastrandrea, M.D., et al. 2011. "The IPCC AR5 Guidance Note on Consistent Treatment of Uncertainties: A Common Approach Across the Working Groups." *Climatic Change* 108: 675–91.

National Academies of Sciences, Engineering, and Medicine (NAS). 2016. *Attribution of Extreme Weather Events in the Context of Climate Change.* Washington, DC: National Academies Press.

Oreskes, Naomi, and Erik Conway. 2010. *Merchants of Doubt: How a Handful of Scientists Obscured the Truth on Issues from Tobacco Smoke to Global Warming.* New York: Bloomsbury.

Sarewitz, Daniel. 2016. "Saving Science." *The New Atlantis* Spring/Summer: 4–40.

—. 2000. "Science and Environmental Policy: An Excess of Objectivity." In *Earth Matters: The Earth Sciences, Philosophy, and the Claims of Community,* ed. Robert Frodeman, 79–88. Hoboken, NJ: Prentice Hall.

Schneider, Stephen, and Kristin Kuntz-Duriseti. 2002. "Uncertainty and Climate Change Policy." In *Climate Change Policy: A Survey,* ed. Stephen Schneider, et al., 53–87. Washington, DC: Island Press.

CHAPTER 7

How Do We Know?

Methods and Tools

Grand Canyon Historic Havasupai Point

FIGURE 7.1

The "scientific method" is usually defined as a process of gaining knowledge that involves observation, forming a hypothesis or theory, making a prediction, performing a test, and analyzing results. This should sound familiar, because this is central to the "constitution of knowledge" as a truth-testing process. Indeed, scientists use norms and methods together with the tools that assemble the body of scientific knowledge.

Of course, scientific methods often take place in laboratories or computers, but they also happen out in the field. My favorite example comes from the scientist, explorer, and soldier, John Wesley Powell (1834–1902) and his colleagues at the US Geological Survey. In 1869, Powell led the first expedition of non-Native American explorers through the Grand Canyon. They wanted to learn practical things about the area and its prospects for settlement. Powell wisely recommended using watersheds to determine political boundaries in the West—advice that was ignored.

He was also fascinated by the question: How did the canyon form? Though he had lost his right arm in the US Civil War, Powell would climb the canyon walls with a pack of scientific instruments. He surveyed the landscape, studied the rocks, and listened to the Native Americans in the area. All the while, he was turning ideas around in his head and modifying them as he gained more clues. In these ways, Powell and later scientists pieced together the story of how the Colorado River acted like a buzz saw cutting down into the uplifting Colorado Plateau.

We call this canyon "Grand" now because of the way science, working together with art, discloses its wonders to us (see Pyne 1999). When you stand on its rim and look down to the Vishnu Basement rocks below, it is science that blows your mind with the realization you are peering back in time two billion years. As your eyes scan the layers of rock you are traveling through millions of years and through deserts, mountains, and oceans. You are also transported in space with the realization that when those ancient rocks were formed, they were on a plate floating on the Earth's molten mantle somewhere far from their current location.

The key here is that in addition to the scientific *method* (as a process of truth-testing), the scientific *methodology* (as a *logos* or way of reasoning) compels us to step outside of our day-to-day, earthbound human perspective. Galileo's telescope symbolizes this shift of perspective even better than Powell climbing out of the canyon (see Arendt 1958). It is true that earlier thinkers like Bruno and Copernicus had used their imaginations to go beyond sense experiences and traditional wisdom. They were able to speculate as if they were floating above their bodies and even beyond the Earth. But the telescope took this raw ingredient of scientific imagination and moved it into sense perception. People could really see as if they had been transported beyond the Earth.

It is now unremarkable to look at a globe or explore satellite images on a computer as if we were floating through space. Though we still stand on Earth, modern science gives us powers to see it and to act on it as if we stood at a remove. This is a revolutionary new way of being and perceiving—to see that movements on Earth are just a special case of the same exterior forces at work on all heavenly bodies. In the same way, the microscope and other tools allow us to zoom into the human body to see cells, proteins, DNA—in other words, to see ourselves as just a special case of the same forces at work on all living bodies.

All this is to say that the methods and tools of the climate sciences push us outside of our usual, limited point of view in order to help us gain knowledge. In addition to testing knowledge-claims, scientific methods and tools can:

* Expand our senses, allowing us to detect phenomena—like GHG concentrations or Ph levels[1]—that would otherwise be imperceptible.
* Quantify otherwise subjective phenomena. Tools that measure temperature put feelings into numbers. We can hardly talk about rising global average temperatures if we can't measure some quantity called "temperature."
* Simulate alternative futures. Computer models help us understand the interactions of our activities with oceans, glaciers, forests, and more under different conditions, helping us to understand various "what if" scenarios.
* Take us back in time. Ice cores have helped us piece together ancient climates.

In this chapter, our goal is to learn more about *how* climate science is done. I organize the methods and tools of the climate sciences into two big categories: climate data and climate models (see Parker 2018). I then focus on climate scenarios (key inputs into models), because these are so important and they form a perfect bridge from this section of the book to the next.

Climate Data

As Figure 7.2 indicates, climate scientists use a wide range of methods and tools to collect various types of observational data. The problem orientation

---

1   A scale of how acid or basic something is, ranging from 0 (very acid—e.g., battery acid) to 7 (neutral—e.g., pure water) to 14 (very basic—e.g., drain cleaner).

framework is useful once again for understanding what all this data is *for*. We need climate data for:

- Detection of trends: filtering signals from the noise to determine where we have a long-term, statistically significant trend in climate.
- Attribution of conditioning factors: examining the causal chains that explain the observed trends.
- Development of projections: creating plausible simulations of climatic conditions in the future based on different choices that we can make in the present (i.e., scenarios).

Global Observing System

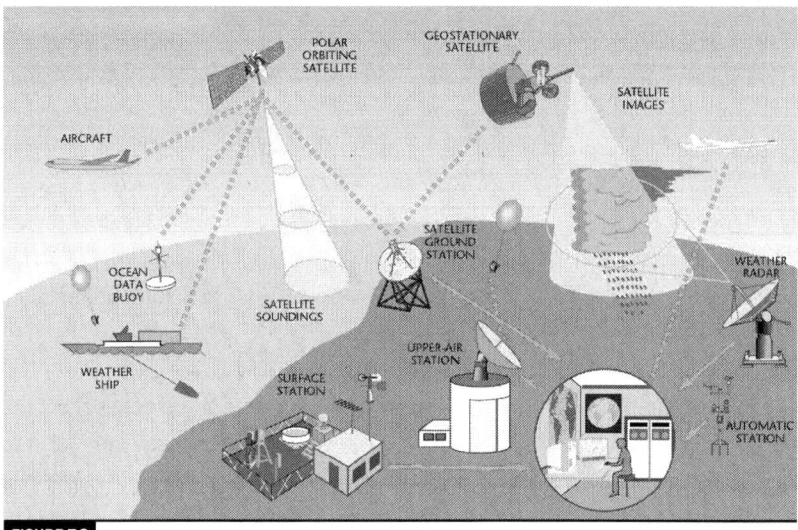

FIGURE 7.2

Following the work of climate philosopher Wendy Parker (2018), we can group climate data into three main kinds.

Station-Based Data Sets

Since the mid-nineteenth century, a global network of land-based observation stations has been collecting data about temperature, precipitation, pressure, and more. Thousands of stations are recording billions of data points. The *constitution* (as a verb) of climate knowledge involves efforts to render all that information into data sets that give us reliable insights into trends. This entails merging observational records, subjecting them to quality control, and homogenizing them (correcting for, e.g., gaps in records when an instrument was offline, differences in altitude, urban versus rural settings, and so on).

Climate scientists have also developed several methods for gridding datasets, that is, providing temperature and precipitation values at points on a grid, where each point is tagged to a spatial region (e.g., 30° N. latitude x 25° W. longitude). The methodological choices behind data usage have fueled claims by climate change deniers that scientists are "manipulating" data. There are important normative questions at stake here about legitimate data practices. However, independent methodologies have all converged around the same trend of warming in the late twentieth and early twenty-first centuries. As noted earlier, this imperative to test claims from multiple angles gives the scientific body of knowledge a unique robustness and credibility.

Though it bends the category of "station-based data," I have to also mention ECHO, a little yellow robot in Antarctica that is helping scientists monitor emperor penguin populations (Dutfield 2022).

ECHO Rover Approaches an Emperor Penguin Colony in Atka Bay, Antarctica

FIGURE 7.3

Remote robotic systems, including drones, can allow scientists to collect data in ways that minimize interference with study populations and eco-systems. Sadly, ECHO has confirmed worries that emperor penguins face a growing threat of extinction due to sea ice loss.

Reanalyses

Of course, there is much more to the climate system than just the land surface where these observation stations are located. There are miles of atmospheric heights and ocean depths that play crucial roles in determining climatic con-ditions. Scientists do have some tools for gathering data directly in these wider dimensions, including instruments affixed to balloons, satellites, and ocean floats. Yet these tools do not provide even, global coverage of the climate system and many of them have only come into use quite recently.

To remedy such gaps, scientists have developed methods of data assim-ilation. These are ways to estimate the three-dimensional state of the atmo-sphere and ocean using observations plus physics-based simulation models. In weather forecasting, the resulting estimate is called an analysis. Climate researchers produce long-term datasets by performing data assimilation for a sequence of past times (e.g., every 12 hours over many decades). This produces a retrospective analysis or *reanalysis*, which provides complete gridded data at regular intervals over a long range of time.

Paleoclimate Reconstructions

Just as there is more spatial depths to the climate system, there are also vast stretches of time that are beyond the reach of direct observational mea-surement. To study ancient climates (paleoclimatology) scientists rely on *proxies*, which are some aspects of nature that can be interpreted to rep-resent climate-related variables. For example, scientists can use oxygen isotope ratios in ice cores or fossils as proxies for temperature, they can use ancient pollen grains buried deep in lake sediments to reconstruct past eco-systems, and they can use the width of tree rings as a proxy for precipitation. With such tools, they can reconstruct past climates.

Here too, the constitution of knowledge involves a host of challenges related to methods, tools, and their norm-governed use. As just one example, consider the difficulties in interpreting tree ring data, which may be influenced not just by temperature and precipitation but also soil quality and other micro-environmental factors that may not be straightforward proxies for the climate. Yet think back to last chapter when we discussed the evaluation of evidence. By drawing from many different kinds of evidence and comparing the findings of different expert groups, the IPCC can draw inferences to conclusions about ancient climates with varying levels of confidence.

Climate Models

I love those realistic train models that capture a scene with great detail: a little train runs over the mountains and into a village where tiny figures are bustling from the post office to the library. Models are representations or simulations of reality at a smaller scale. When they represent open systems like a mountain village or the climate, they are necessarily incomplete simplifications. We know, for example, that no matter how finely crafted the figures in that scene are, they do not have tiny working hearts, living cells, or active minds. To capture every functional detail of something would require a map the size of what is mapped!

Computer models are powerful tools for studying the climate system. Digital computers make it possible to simulate the large-scale movement of mass, heat, carbon, air, moisture, ice, and other components of the climate system. They use complex mathematical equations derived from scientists' understanding of the physical laws that govern the climate system. Models clearly express the worldview of the climate sciences (and all modern science). They assume the climate system is governed by natural laws that can be expressed in mathematical equations (e.g., about fluid motion, thermodynamics, etc.). Again, this is the methodology associated with Descartes and the sixteenth- and seventeenth-century scientific revolution: imagine that God created the universe and then left it alone to operate in accordance with the laws God had established. In a way, scientists too create their little simulated worlds inside of the models and then let them run to see what emerges from the interaction of all those complex equations.

Models allow scientists to further understand the climate. Models are especially useful for:

* Testing causal explanations for how different conditions and processes bring about phenomena of interest, like global warming; and

- Making long-term conditional predictions known as *projections*, which are predictions of what future climate would be like under particular conditions known as scenarios (more on those below).

In this section, we'll survey some main types of climate models and discuss their evaluation.

Types of Climate Models

Scientists often speak of a hierarchy of climate models ranging from the simple to the complex. As Figure 7.4 illustrates, more complex models have been developed over time as more powerful computers have been able to include more climate system components and processes. Complexity also increases as models represent more spatial dimensions with greater resolution[2] and as they increase the "timestep," that is, how often they calculate the state of the climate system (see McSweeney and Hausfather 2018).

Here's one way to arrange the hierarchy of models starting at the simplest end:

- Energy Balance Models can be run on a few lines of code because they only consider the balance between the sun's energy entering the atmosphere and the heat radiated back to space. Surface temperature is the only climate variable calculated.

- General Circulation Models or Global Climate Models (GCMs) simulate the physics of the climate by capturing flows of air and water through the heights and depths of the atmosphere and ocean. More complex versions are known as coupled atmosphere–ocean general circulation models (AOGCMs), because they link together multiple models to get a more comprehensive representation of the climate system. Regional Climate Models are like GCMs for a specific area, which allows them to be run more quickly and/or with greater resolution.

- Earth System Models (ESMs) improve on GCMs by including more complex representations of vegetation, ocean ecology, and biogeochemical cycles.

---

2 I.e., greater detail, more accuracy.

* Finally, Integrated Assessment Models (IAMs) add yet more complexity by including social factors. Scientists typically use mathematical equations to represent unchanging physical laws. With social factors, however, this is not possible—the social sciences are far less able to find laws comparable to those in the natural sciences—laws about, for example, how the economy will develop, how land and energy policies will change over time, or how politics and culture will evolve. This means that there are many debates in the climate science community about what theories and assumptions to use when modeling human elements of the dynamic climate system.

The Evolution of Climate Models

For decades scientists have been using **mathematical models** to help us learn more about the Earth's climate. Known as climate models, they are driven by the fundamental physics of the atmosphere and oceans, and the cycling of chemicals between living things and their environment. Over time they have increased in complexity, as separate components have merged to form coupled systems.

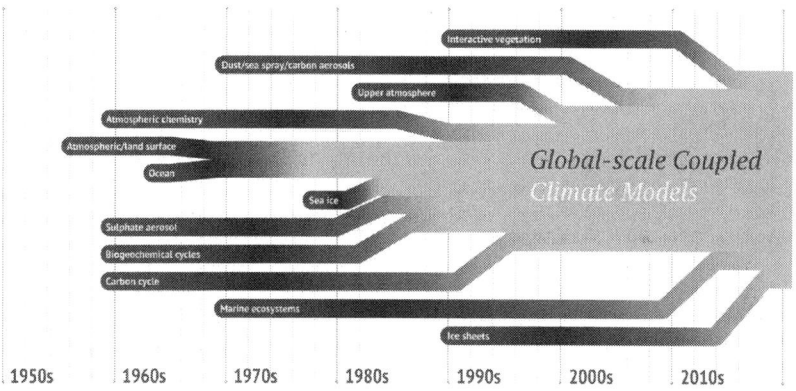

Note: There were some very simplified models before the dates mentioned.

FIGURE 7.4

The evolution of models can improve our understanding and, in some cases, give us further hope and motivation for action. For example, earlier GCM models used simplifying assumptions that held atmospheric concentrations of GHGs constant across a range of scenarios. With their ability to capture more complex biogeochemical cycles, later generations of ESM models move us past this assumption. Most ESM models indicate that once we reach net zero GHG emissions, CO_2 concentrations in the atmosphere start to fall quickly, because land and ocean sinks absorb more carbon than we are emitting into the atmosphere. In other words, the climate science community used to believe that a certain amount of warming was "locked in" even after emissions are eliminated. It now appears (though recall that uncertainty is always present) that is likely an inaccurate picture of how the climate system will behave.

Evaluating Climate Models

In some way, climate models are always "wrong," because they are approximations of a chaotic, complex, open system. Yet this does not diminish their potential value. Remember, the reason for models is our lack of access to the phenomena we want to study. For example, we just don't have observations of the future! So, when evaluating climate models, the standard is not an exact match to every detail of reality. Rather, evaluations look to the respects in which and degrees to which models are similar to the climate system (Lloyd 2010). There is that key idea again—how is the model *like* the world? It's another kind of analogical reasoning.

Some important tests for models are measures of fit between model outputs and observational data. In some cases, the IPCC has very high confidence about the ability of models to reproduce features of the climate system. Most importantly, this includes the global-scale annual mean surface temperature increase across the twentieth and early twenty-first centuries. In many other cases, however, it is hard to gauge whether model-data comparisons provide evidence for a model's similarity to the climate system or its usefulness for various purposes. A poor fit between observed data and model outputs might be because the data are flawed, the model contains erroneous assumptions, the initial conditions of the simulation were inaccurate, or other factors.

If a model's outputs match observations, it is tempting to say that the model has been "verified." But this is to commit the logical fallacy of "affirming the consequent"[3] (see Oreskes et al. 1994). Consider the example: "If it snows tomorrow, I will wear my boots." When tomorrow comes, you see that I am wearing boots and conclude that it is snowing. But it may not be snowing—maybe I just lost my other shoes. Similarly: "If the model is true,

---

3   Affirming the consequent is a mistaken form of reasoning using a "hypothetical" statement—one of the form "If X then Y." The Y-part is called the consequent. The mistake is to use a hypothetical statement, together with the affirmation of the consequent, to establish the X-part, the antecedent: If X then Y; Y; therefore X.

then it will match observed results." When the model matches observations, though, we cannot conclude it has been verified (meaning proof of truth). The model may be right but for the wrong reasons (maybe it got lucky), meaning it may not be right next time. On the other hand, especially if the model gets it right several times, that adds some likelihood that the model is right. But this is still all just a matter of probability. It's all part of the provisional nature of the constitution of knowledge and the unavoidability of uncertainty in science. So, it is better to speak of "confirmation" than "verification."[4] The greater the number and diversity of confirming observations, the more support we have for the probability that the model is not flawed.

Climate Scenarios

We can think of models like machines that process inputs to produce outputs. Inside of the machines, as we have seen, are mathematical equations and other "if-then" operators. Hopefully you have a sense of just how complex this can get. Now, though, let's close the machine and treat it as a black box to focus on the inputs. After all, as the old saying goes about computing: garbage in, garbage out. Particularly when it comes to the output of projections, their quality can only be as high as the quality of inputs we feed into the models.

The main inputs into climate models are "forcings," which change the climate system in some way. Forcings are like kicks to the system. When scientists run models, they are usually interested in how they behave under the influence of a different mix of forcings. Some forcings are beyond human influence like wobbles in the Earth's orbit, solar cycles, and volcanic eruptions. Others, though, are anthropogenic, like deforestation and GHG emissions.

Humans (at least some humans) have become major forcings in the climate system. This human element of a changing climate, though, is also the most uncertain and unpredictable. Try to imagine what the future will be like in 80 years. Before you get too confident, think about how successful someone in 1920 would have been in guessing what would transpire by 2000. Human action is boundless—causing cascading consequences that constantly shuffle all the pieces.

---

4   "Verification" means showing *true*, so it is slightly misleading; if the best science can do is show a hypothetical model is *probable* then it might be better to speak of its processes as "confirmation"—suggesting increase in probability of truth, of likelihood.

RCPs and SSPs

So how do we take all of this uncertainty and complexity and turn it into inputs for climate models? One key way is via *emissions scenarios*, which are potential pathways toward different future concentrations of GHGs in the atmosphere. Keep in mind the main goals are to avoid "dangerous anthropogenic interference" while improving human well-being. Scenarios give us tools for thinking about how we may—or may not—achieve those goals.

Having a small number of standard scenarios as inputs also allows the climate science community to streamline their work and compare results across hundreds of types of models. The IPCC has been using scenarios since 1992. In 2007, a process started to update the scenarios. Beginning with the AR5 (2013–14), the IPCC has used two new kinds of scenarios to develop projections and shape its advice for policymakers.

1. *Representative Concentration Pathways (RCPs).* These describe different levels of GHGs and other radiative forcings that might happen in the future. Climate researchers developed four such pathways, spanning a range of forcing in 2100. Each was numbered for the radiative forcing it described in terms of watts per square meter: 2.6, 4.5, 6.0, and 8.5. In other words, the four RCPs are model inputs that sketch possible future trajectories (pathways) of emissions ranging from low to high. These, in turn, result in a range of atmospheric concentrations of GHGs by 2100. This allows scientists to use their models to study how the various parts of the climate system might react to these forcings. Of course, they are particularly interested in how the scenarios relate to the 1.5°C and 2°C targets, but they also study a wide range of climate variables. RCPs do not include any narrative elements—unlike what's described just below, that is, they contain only numbers, no interpretive prose.

2. *Shared Socioeconomic Pathways (SSPs).* By contrast, SSPs (which were first used in the IPCC AR6, 2021–22), center on plausible stories about how society might change over the twenty-first century (see Hausfather 2018). They are designed to work in tandem with RCPs. As Figure 7.5 shows, there are five SSPs that can each be paired with the different RCPs. Each of the resulting scenarios tells a different story about how global society will change over the twenty-first

century. The stories are based on assumptions about plausible developments in geopolitics, population, GDP, energy, climate, land use policies, and technological innovation.

Global CO_2 Emissions (GtCO_2) for All IAM Runs in the SSP Database Separated Out by SSP

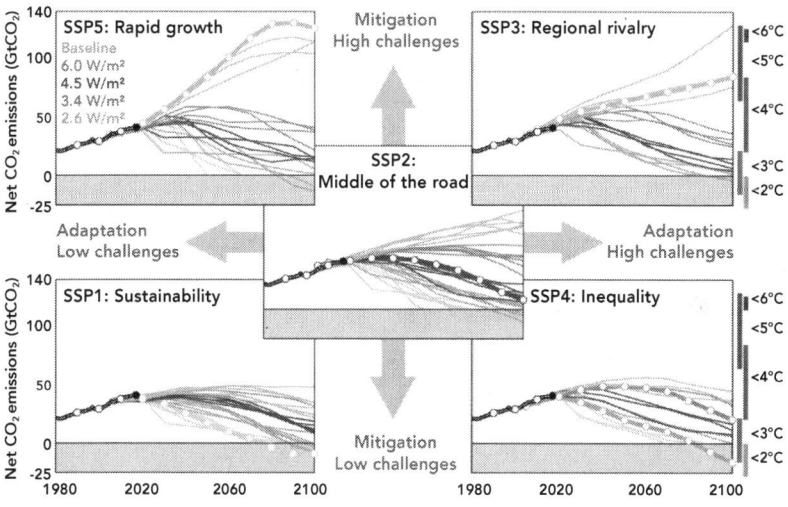

FIGURE 7.5

Models and scenarios will continuously need to be updated as we learn more and as conditions change under the influence of unfolding human actions. In the words of Yogi Berra, "the future ain't what it used to be." These updates are a crucial form of the progressive, self-correcting nature of science. Without examining our assumptions and discarding the outdated ones, the constitution of knowledge loses its integrity and credibility.

The most important example here pertains to RCP 8.5 (see Pielke and Ritchie 2021). In the AR5 (2013–14), the IPCC labeled RCP 8.5 the "business-as-usual" scenario, meaning it was the most likely future in the absence of policy interventions. However, by AR6 (2021–22), the IPCC shifted its assessment, writing that "the likelihood of high emissions scenarios such as RCP 8.5 or SSP5–8.5 is considered low in light of recent developments in the energy sector" (IPCC AR6 WG1, p. 236). One main driver of this shift is that past assumptions about high global coal use were proving wrong, as renewable energy prices dropped faster than expected, prompting major energy transitions.

Scenarios are inevitably full of assumptions, because there is no way to create a narrative as richly detailed as real life. So, examining those assumptions is an important part of ensuring that scenarios help us imagine better futures. As noted in Chapter 3, justice needs to be central to these visions of the future. According to a team of researchers from India, the scenarios used by the IPCC fail in this regard. For changes from 2020 to 2050, the scenarios frame "success" (limiting warming to 1.5°C) in ways that "project existing global inequities far into the future" (Kanitkar et al. 2022). For example, growth in per capita consumption of goods and services in one scenario for Sub-Saharan Africa grows from $2,000 to just $4,000, while in North America the growth is from $51,000 to $79,000. Meanwhile, fossil fuel consumption in the Global North remains high in 2050, in fact, much higher than fossil fuel consumption in the Global South. Is this justice? The authors argue that it is not—the Global North does not live up to its responsibilities and the development needs of the Global South are disregarded.

Conclusion

The methods and tools of climate science work in tandem with norms to constitute the growing, dynamic body of knowledge and uncertainty. Climate data is collected from a network of tools and proxies and subjected to a variety of methods for turning raw information into useful insights about trends, conditioning factors, and projections. Climate models, which are run on increasingly powerful computers, allow us to test hypotheses, explore "what if" questions, and guide further research. And climate scenarios exercise our imaginations and give us projections to inform the development of policy.

As noted earlier, the remainder of the book will bring the social sciences like economics and the humanities back into the conversation. Scenarios like the SSPs are a perfect bridge, because they draw from these subjects to create stories. The humanities and social sciences study people as *biographical* beings—creatures who make and are made by stories—and not just *biological* beings. Climate scenarios are fascinating points of contact between the non-narrative rationality of science and the humanistic realm of meaning and action. In other words, they bring us into the last section of the book on climate politics, policy, and ethics.

Activities and Questions

1. Find a climate change story in the news with an alarming headline based on climate research. Trace back to the research article being discussed and see if you can find out what RCP or SSP the authors used. Is there a connection between such research choices and the kinds of media coverage and headlines that reach the general public?

2. Find out if there are climate data-collecting tools in your locality. Who operates them and how are the data used?

3. Check out the University Corporation for Atmospheric Research (UCAR) Center for Science Education's online tool called "The Very Simple Climate Model." You can manipulate a few forcing variables to see how outputs change.

4. Which of the SSPs do you think is most plausible and why? Try writing your own SSP and imagine what kind of projections the models would produce if your scenario was used as an input.

References

Arendt, Hannah. 1958. *The Human Condition*. Chicago: University of Chicago Press.

Dutfield, Scott. 2022. "Meet the Robot Keeping an Eye on Emperor Penguins in Antarctica." Live Science, May 19. https://www.livescience.com/robot-in-antarctica-watches-over-penguin-population.

Hausfather, Zeke. 2022. "The Case for Cautious Climate Hope." University of Minnesota, 29th Kuehnast Lecture. https://www.youtube.com/watch?v=XgosvsTWC-k.

—. 2018. "Explainer How 'Shared Socioeconomic Pathways' Explore Future Climate Change." Carbon Brief, April 19. https://www.carbonbrief.org/explainer-how-shared-socioeconomic-pathways-explore-future-climate-change/.

IPCC. 2021. *Climate Change 2021: The Physical Science Basis. Contribution of Working Group I to the Sixth Assessment Report of the Intergovernmental Panel on Climate Change.* Cambridge: Cambridge University Press.

Kanitkar, T., A. Mythri, and T. Jayaraman. (2022, November 3). *Equity Assessment of Global Mitigation Pathways in the IPCC Sixth Assessment Report.* https://doi. org/10.31219/osf.io/p46ty.

Lloyd, Elisabeth. 2010. "Confirmation and Robustness of Climate Models." *Philosophy of Science* 77 (5): 971–84.

McSweeney, Robert, and Zeke Hausfather. 2018. "Q&A: How Do Climate Models Work?" Carbon Brief, January 15. https://www.carbonbrief.org/qa-how-do-climate-models-work/.

Oreskes, Naomi, et al. 1994. "Verification, Validation, and Confirmation of Numerical Models in the Earth Sciences." *Science* 263 (5147): 641–46.

Parker, Wendy. 2018. "Climate Science." In *The Stanford Encyclopedia of Philosophy,* ed. Edward N. Zalta. https://plato.stanford.edu/archives/sum2018/entries/climate-science.

Pielke, Roger, and Justin Ritchie. 2021. "How Climate Scenarios Lost Touch with Reality." *Issues in Science and Technology* 37, no. 4 (Summer): 74–83.

Pyne, Stephen. 1999. *How the Canyon Became Grand: A Short History.* New York: Penguin.

PART 3

Climate Politics, Ethics, and Policy

CHAPTER 8

Climate Politics

East African Crude Oil Pipeline

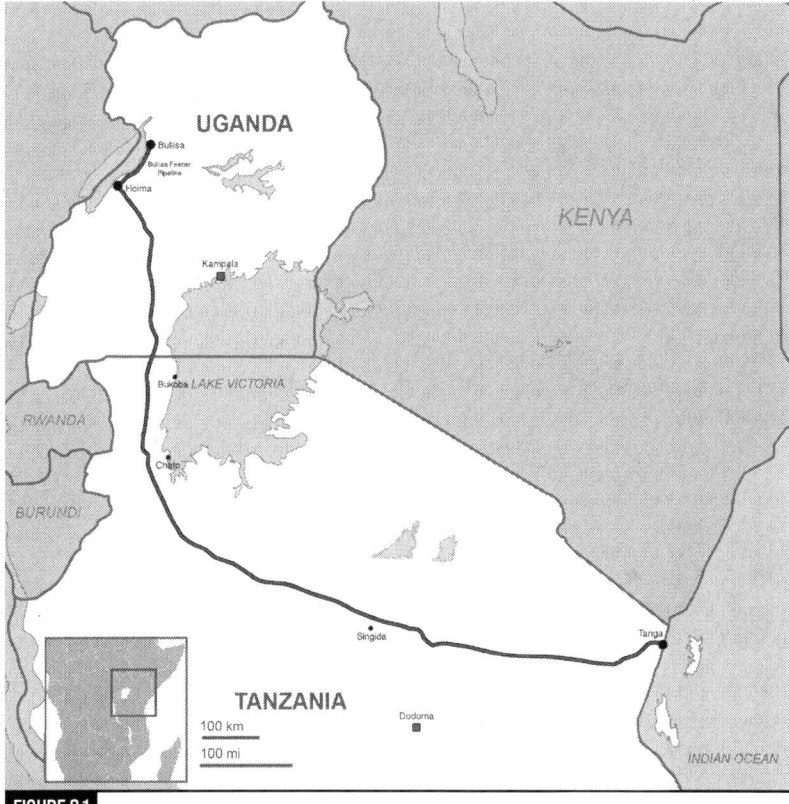

UGANDA
Bulisa
Bulisa Feeder Pipeline
Hoima
Kampala
KENYA
Bukoba LAKE VICTORIA
RWANDA
Chato
BURUNDI
Singida
Tanga
TANZANIA
Dodoma
100 km
100 mi
INDIAN OCEAN

FIGURE 8.1

U nderstanding the natural sciences is a necessary part of climate literacy, but it is not sufficient. Recall the IPCC quote from the introduction: sciences provide "essential information and evidence for decisions" but "at the same time, such decisions are value judgments." Science and values come together in climate politics, ethics, and policy. This final section of the book offers tools for thinking through these tangles so that you can help society make good decisions.

Let's start with a case study for a concrete illustration of climate politics. In April 2021, a signing ceremony concluded the final agreements to launch the Lake Albert Development Project in Africa. The parties involved were the Presidents of Uganda and Tanzania, the Ugandan and Tanzanian national oil companies, the French company TotalEnergies SE, and the China National Offshore Oil Corporation. The project includes oil development in Uganda and the construction of the East African Crude Oil Pipeline (EACOP). The pipeline is expected to deliver about 230,000 barrels/day to the Tanzanian port city of Tanga (see Figure 8.1).

The project sparked lots of controversy (Olewe 2022). The European Union Parliament issued a resolution denouncing EACOP. They called for an end to drilling in protected areas, fair compensation for people displaced from their land by the pipeline, an end to human rights violations of protestors, and further study of different routes and alternative projects to develop renewable energy instead of oil. Environmental advocates argued that the IPCC reports have one conclusion: no more fossil fuel production. Burning the oil carried by EACOP is estimated to create 34 million tons of CO_2 annually. Others argued that the profits from the project will not be fairly distributed.

Many leaders in Uganda and Tanzania replied that the EU was being hypocritical. In 2021, both African nations got over 80 percent of their energy from renewables, whereas that figure for the EU was just 22 percent. The entire continent of Africa has only emitted 3 percent of GHGs, compared with 17 percent from the EU. Plus, all the oil in the EACOP would increase global GHG emissions by less than 0.05 percent. Meanwhile, the project would generate billions in annual revenues. Proponents argued that African nations have the right to use their fossil fuels for development, just as wealthy nations have done for hundreds of years. If wealthy nations want Africa to accelerate its energy transition, then they should provide more climate financing.

What is the right decision? Imagine that we had a Climate Computer powered by artificial intelligence. It is programmed to achieve our two main goals: a) Decrease climate dangers (safety) and b) Boost development (well-being). Could such a machine render politics into a form of calculation? Could it land upon some "objectively" right decision in situations like EACOP? I don't think it could. There is no way to turn human affairs—the stuff of politics—into a set of equations that could be "solved."

Consider all of the values that the programmers would have to build into the computer. How are the two main goals coded—what do "danger" and "development" mean? What evidence would the computer consider? How would it handle uncertainties about, say, pipeline leaks, emissions, or revenue? Even if the computer could predict such things (an impossibility for open systems), how would it balance the resulting costs and benefits? How would it reckon with things that are costs for some but benefits for others? What is the "objectively" right scheme of financing and revenue distribution?

In other words, the computer would be doing politics by other means. It would be a form of technocracy—rule by experts—because decisions would be determined by programmers. But why should they make such values judgments? Shouldn't that be left to the people impacted by the decision? And do you think people would passively submit to the edicts rendered on high by the Great Climate Computer?

There is no avoiding politics, the activities of group decision-making that determine who gets what, when, how, and why (Lasswell 1936). Collective decisions have to be made one way or another, which means the power to decide has to be distributed one way or another. Yes, politics is driven by strong emotions (fear, resentment, greed, patriotism, etc.) and clashing interests that are not always rational. Nonetheless, we can step back to rationally analyze the political process. For climate literacy, we need the *human and social* sciences as much as the *natural* sciences. These are not exact sciences, to be sure, but they offer tools to organize thinking and guide action.

In this chapter, we'll first build a bridge from the previous section of the book by critiquing a flawed model of climate politics that pictures political reasoning as just the application of science. Then, we'll examine a better model that pictures political reasoning as the ability to see and resolve paradoxes. Next, we'll add to this skill of paradox-thinking another tool: the social process. To conclude, we'll return to climate as a hyperobject to think about its radical political implications: if politics is the art of the possible, what is the scope of possibility in a world being transformed by accelerating changes and massive flows of energy?

The Linear Model of Climate Science-Politics

Most approaches to climate literacy hope that knowledge will clear away the clouds of ignorance and conflict that so often shroud politics. Science will dispel disagreement and compel consensus around the right decision. This model, like the Climate Computer, pictures politics as just the application of science (see Sarewicz 1996). It is often called "the linear model."

The Linear Model

Science ➡ Knowledge ➡ Political Consensus ➡ Wise Policy Action

FIGURE 8.2

There is an important kernel of truth to this model. Good decisions and sound policymaking should be based on knowledge, evidence, and reason—at least that is the moral conviction behind democratic political regimes. And the constitution of knowledge should serve as the process for acquiring knowledge that is independent of any political ideology or religious worldview. Granted, the Climate Computer shows the limits of trying to turn politics into calculation. But imagine the other extreme where there are, say, Liberal/Democrat and Conservative/Republican "scientists" and "science" is just another name for partisan battles. That is the pitfall of false equivalence that would come from an extreme politicization of science. We need to be vigilant about threats to the autonomy, integrity, and credibility of the constitution of knowledge. Remember, the threats posed by "merchants of doubt" and epistemic "trolls" (Oreskes and Conway 2010; Rauch 2021).

And yet even if we could eliminate all the "trolls" (and could we do this without violating free speech ideals?), the linear model would still be flawed. The challenges of reaching wise action extend beyond the bad faith activities of the carbon industrial complex. In the EACOP example, both sides are advancing legitimate arguments based on evidence. To reduce the controversy to a clash between those with "real" science and those with "junk" science is to fall into the false binary pitfall. Figure 8.3 shows how there are complicating factors at nearly every step in the linear model.

In the constitution (as a verb) of knowledge, we saw that there are inevitably uncertainties, disagreement, and various interpretations and judgments around methods and tools. The resulting knowledges and uncer-

Complicating Factors in the Linear Model

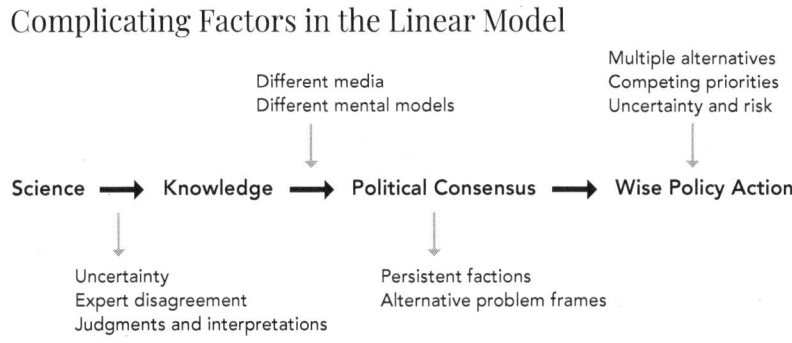

Different media
Different mental models

Multiple alternatives
Competing priorities
Uncertainty and risk

Science ⟶ Knowledge ⟶ Political Consensus ⟶ Wise Policy Action

Uncertainty
Expert disagreement
Judgments and interpretations

Persistent factions
Alternative problem frames

FIGURE 8.3

tainties (both plural!) reach political processes through a variety of media and are filtered through different mental models. Yes, some of this involves disinformation campaigns, but much of the "spin" comes from reasonable differences in points of view. This means that political factions can assemble different problem frames. As a result, deciding on the wise action will require slogging through debates on the best available alternatives, the right ranking of priorities, and the proper balancing of risks, costs, and benefits. That's why it is important to develop our skills for thinking about how those debates are—and should be—designed and how we can effectively intervene.

The limits of the linear model are rooted in the limitations of the modern natural sciences to:

- Articulate significance and meaning
- Distinguish qualitative differences
- Formulate problems
- Guide action

The modern scientific worldview makes nature appear as modifiable but does not say which modifications to make. Another oil pipeline? More nuclear power? Lab-grown meat? Mirrors in space? It all raises the political and ethical question: What should we do?

Politics and Paradox

So, politics is not just applied science. What is it all about, then? Let's first make some basic points about politics before focusing on political reasoning as the art of paradox.

What Is Politics?

Most people define politics with one word: power. We think of a politician cutting deals to retain his position and maybe even line his pockets from wealthy patrons. Or we think of the *Realpolitik* described by the ancient Athenian historian Thucydides: "The standard of justice depends on the equality of power to compel and that in fact the strong do what they have the power to do and the weak accept what they have to accept." Might makes right.

There is no denying the truth of this portrait. And yet it is not the whole picture and does not capture what is most fundamental about politics. Note how there is an implicit critique of those snapshots of raw power in the previous paragraph. Yes, we admit that this is often how things go, but that's not how they *should* go. Further, if politics is all about power, then it would just be another name for violence. Terror and tyranny would reign everywhere.

Politics is more fundamental than power or factions. It is about human plurality: each person is unique—with their own values, mental model, and experiences. And yet we can only express and realize our uniqueness by living together in common. At its basic level, politics is what the philosopher Hannah Arendt (1958) called the "space of appearance."[1] It is the human world that testifies to the presence of others—where we shape and are shaped by others and where we struggle to be recognized.

The ancient Greeks understood politics as the sphere of human freedom, because it is here that we rise above the mere necessities of labor and the household realm to express and pursue our ideals. Aristotle described humans as the political animal (*zoon politikon*) not just because we have families or because we are social creatures. Ants and bees are also social. What is unique about humans is that we are capable of deciding our common affairs through speech, reason, and persuasion. We can freely decide our common direction rather than being compelled by the impulses

---

1   Arendt means "where I appear to others as others appear to me, where men exist not merely like other living or inanimate things, but to make their appearance explicitly." In this "space" we can act together for the public good (1958, 198–99).

of instinct or the brute force of violence. This is why the concept of the *common interest* is so vital for politics. Indeed, politics is the process of rising above brute power of the strong over the weak to seek the common good for all.

Building on this idea, John Locke argued that a political society is defined by an agreement to live by a set of laws established for the common good. It is that ideal that informs our critique of raw power. The politician should serve the people and not just his own interests. The warmongering state should respect the right of other peoples to exist. The ideal is the reverse formulation "Right makes might." Power should flow only from just laws. But what is justice and what are just laws? What is the common interest? These questions are at the heart of the human and social sciences of climate change, and so they will be recurring themes for the remainder of the book.

That is, we can also think of politics as a kind of science, a form of reasoning that we are practicing as part of climate literacy. This science asks both descriptive and normative questions about how people *do* and *should* conduct their lives and organize their societies. There are many different ways to organize political life, each with its own overarching aim that defines what a good society is. For example, we can imagine:

* Rule by the techno-scientific elite in pursuit of the means of survival and comfort
* Rule by the morally wise in pursuit of justice and the good
* Rule by the wealthy in pursuit of riches
* Rule by the patriotic in pursuit of honor, glory, and nationalism
* Rule by the tyrant in pursuit of absolute control and security
* Rule by the righteous in pursuit of salvation for the afterlife
* Rule by the people (*demos*) in pursuit of freedom and equality

Political science asks: What kind of regime is best? Which is truly good and just? What laws, conventions, and customs are most conducive to human flourishing or to what we have been calling well-being or development? What regime is most conducive to multispecies justice? Aristotle called politics the "master science," because it establishes the place of all the other sciences in accordance with these larger visions of the just and the good society. Politics determines what kinds of climate sciences are funded, how much money they receive, and how much influence they have in decision-making processes. So, it is vital to think about what is really going on when people engage in climate politics. I think it is helpful to think in terms of paradoxes.

Political Reasoning and Paradoxes

We have already been practicing political reasoning by asking that basic question, "What is it?" Now we can explore this with more care, because it is so key to understanding and participating in climate politics.

Stated abstractly, a paradox is a situation that is both A and not-A. Figure 8.4 illustrates this with the classic duck-rabbit paradox. What do you see?

The Duck-Rabbit Paradox

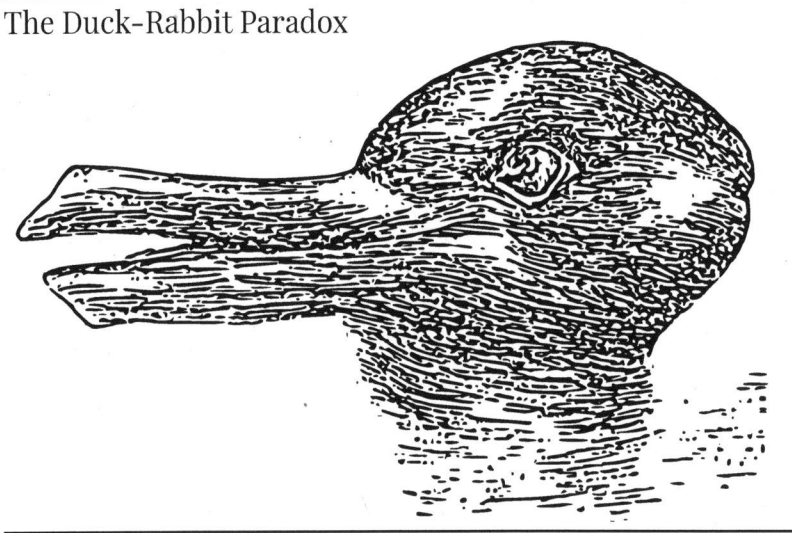

FIGURE 8.4

In quantum mechanics, wave-particle duality offers another example. Every quantum entity can be described as a particle (A) or a wave (not-A). Two contradictory pictures of reality both explain the phenomenon of light. But we don't need such exotic examples—paradoxes are all around. The EACOP pipeline, for example, is an economic boom and a climate bust. It is the right thing to do and the wrong thing to do. What is it?

The political scientist Deborah Stone (1998) argues that it is helpful to see politics as "strategically crafted arguments designed to create paradoxes and resolve them in a particular direction" (4). As we've noted, climate politics is often about what a situation is *like* or what it *is*. Stone puts it this way:

"Political reasoning is reasoning by metaphor and analogy. It is trying to get others to see a situation *as* one thing rather than another" (6).

The history of environmental politics is all about the emergence and contestation of paradoxes. Consider some examples:

- Swamps that are a waste or wetlands that provide essential ecosystem services
- A pristine wilderness developed or humanized ecosystems stolen from Indigenous peoples
- Ornamental plants or invasive species
- Whales as resource or as endangered species
- Storm victims or climate refugees
- And if they are climate refugees, are they intruders or our fellow Earthlings? Who are "they" and who are "we"?

Scientific evidence is often insufficient to resolve a paradox. Mental models largely determine which way a person or a faction will see the situation (thus the kinds of scientific evidence they will focus on). The political scientist John Dryzek (2005) calls mental models "discourses," or shared ways of apprehending and making sense of the world. Discourses are based on assumptions, values, and judgments through which evidence is filtered. This makes politics all about contests over meaning. Importantly, discourses are tangled up with power, and often one discourse becomes the establishment position. That is, a discourse becomes the mainstream view. The UNFCCC's predominant climate change problem framework first introduced in Chapter 3 is the most important example. This can be a double-edged sword. On one hand, if the mainstream discourse is sound, it can lead toward consensus around wise decisions. On the other hand, if the mainstream discourse is faulty, it can be difficult to overturn it.

Of course, we have to keep the pitfall of false equivalencies in mind when we talk about politics in terms of paradoxes. Just because we can see a situation as A and not-A does not mean that both views are equally true. Anthropogenic climate change is happening (A) even if some insist that it is a hoax (not-A). Some paradoxes are resolved through reasoned dialogue and the workings of the constitution of knowledge. (We might call this a meta-paradox: politics both *is* and *is not* about paradoxes.)

The Social Process

Analyzing discourses, arguments, problem-frames, and paradoxes is critical to climate literacy. But these are not just words floating in air. Or, to switch metaphors, the contests over ideas and meaning are not occurring in a vacuum. Speech comes from people who are situated in various social institutions that have certain kinds of resources, power, and interests. In other words, we also need tools for analyzing the material and social contexts of politics. How can we organize our thoughts about the concrete flesh and bones of political actors?

Like the problem orientation, the social process is a tool from the policy sciences (Clark 2002). It doesn't make politics easy. But it does serve as a checklist to help us get a clear picture of the politics swirling around any given decision. It's a useful tool for surveying the landscape and pinpointing your questions, so that you can make sense of the context and understand how you can get involved. Remember, there is a near infinite amount of information we could seek about any given problem. The social process helps us to focus on the *relevant* knowledge so that you don't miss something important in your analysis.

The Main Elements of the Social Process

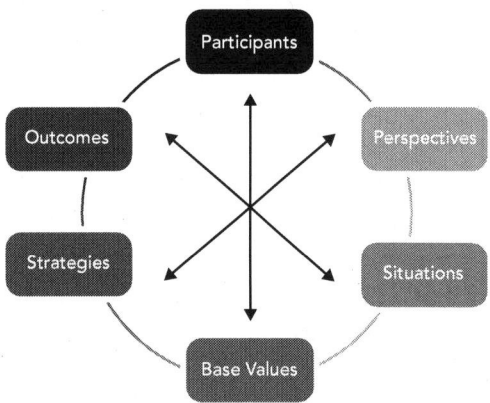

FIGURE 8.5

Figure 8.5 lists the main elements of the social process. This tool gets you thinking about politics as a process where participants are seeking values that they think will leave them better off. They do this through social institutions, and this process has outcomes and long-term effects on people and the environment. By applying the social process, you can make a manageable map of a complex situation. Each situation is unique, but these categories should guide you in gleaning relevant information no matter what the context.

Illustrating the Social Process

From 2011 through 2015, I was involved in the politics of fracking in Denton, Texas (Briggle 2015). Fracking—or hydraulic fracturing—is a suite of techniques for extracting oil and gas from unconventional shale reserves. At that time, Denton was a fast-growing city atop the Barnett Shale, which was a fast-growing site of gas well development. There were over 200 gas wells in the city limits, with many more planned. This led to debates about what the city's policies should be with regard to fracking, which in turn were grounded in different views of what makes a good and just city. I'll draw from my experiences to illustrate the seven categories of the social process. I can't give a full accounting of the fracking case, but I hope to offer enough to give you a sense of how to use this tool.

1. *Participants*: What individuals, groups, and institutions are participating? Who should be involved but is not? Who is demanding to participate? You can classify participants in many ways. For example, individuals might be authorities, experts, or stakeholders and institutions might be local, state, national, or transnational or private, public, or non-governmental.

   In this case, key participants were city council, neighborhood groups, local media, environmental NGOs like Earthworks, mineral owners (on the same property, different people could own the surface and sub-surface rights), members of the gas well development industry, and state politicians and regulatory agencies. City attorneys also played a pivotal role, because much of the debate swirled around jurisdiction: what authority did the city have to regulate fracking?

2. *Perspectives*: How are participants framing the problem? What are their demands (in terms of values, procedures, and outcomes)? What are their expectations and assumptions? And how do they identify (how do they see themselves)? Identity is essential here—how do

participants define "us" and "them"? What narratives, symbols, or myths inform their identities?

The City of Denton framed the fracking case as a land use issue, but the State of Texas framed it as a mineral development issue. It was a paradox. Was it about safe neighborhoods or the state economy? Even the same terms like "private property" were seen in different ways. Whose property—the homeowner or the mineral owner? So much of the politics swirled around identity: "fracking is intruding on *our city*" or "*we Texans* are all about oil and gas."

3. *Situations*: What are the relevant contexts in which the participants interact? These can be ecological (natural), historical/temporal, and institutional. How do geographic and natural resource factors shape the situation? How is power concentrated or distributed? Pay attention to how participants pick and choose certain situations as significant. For example, how do things shift when the situation is described in terms of colonialism?

   Important ecological concerns in the fracking case related to water scarcity and air pollution. The main temporal aspect had to do with the rapid growth of both fracking and residential development. Indeed, it was all sparked by the fact that a booming city just happened to lie atop a major gas field. Big real estate and planning decisions were in the works and hinged on fracking policies. The balance of power between city and state officials was the key situation for the legal and political context. For me, this also meant navigating my responsibilities as a locally engaged academic working at a state institution.

4. *Base Values*: The social process identifies eight values: power (to influence or make decisions), enlightenment (science, knowledge), wealth (money), well-being (health and happiness), skill (special abilities), affection (relationships), respect (standing), and rectitude (ethical standards). The values can be put into two categories depending on the situation. They can be something participants already have as *resources* to seek more values (e.g., they already have wealth and are using it to influence politics). And they can be seen as *goals* or values toward which they are aiming (e.g., they want to have more wealth and are engaged in politics to pursue that goal).

In the fracking case, the state of Texas and mineral owners had more money and power as resource values than the City of Denton and surface owners. Those opposed to fracking emphasized well-being, especially health and neighborhood safety, as their goal value. The proponents of fracking had relationships with key state lawmakers, whereas the grassroots opposition to fracking utilized community organizing skills to develop relationships that were vital for mobilizing activism and voting. Both sides utilized science (enlightenment) in making arguments as well as rhetorical and media skills.

5. *Strategies*: How do participants use their resource values to try to achieve their goal values? What strategies work? Which ones are ethical? Which are missing or under-utilized? You can classify strategies in many ways. For example, diplomatic strategies involve negotiation among elites, whereas other strategies target communication to wider audiences. There are economic, legal, activist, and even military strategies.

In Denton, both sides used strategies to target the wider public as well as decision makers (elites with power). For example, the group I worked with created newspaper ads and crafted reports for City Council. The proponents of fracking used messaging around energy as a national security and economic priority. Our group did not frame fracking in terms of climate change, because it was too partisan in Texas. Rather, the winning strategy was to focus on something everyone could embrace: safe and healthy neighborhoods. Our message was that heavy industry does not belong next to homes.

6. *Outcomes*: The outcomes of politics can be seen as changes in the distribution of values. In the interactions between participants, who gained what and who lost what? This is politics: the shaping and sharing of values through social processes. Admittedly, that is a dull description of an often brutal reality! But it is pretty accurate.

One outcome in the fracking case was a revised city ordinance in 2013 designed to strike a better balance between the competing interests. It was supposed to keep fracking at least 1,000 feet from homes. Yet for a variety of complex legal reasons (e.g., vested rights), fracking continued to happen as little as 250 feet away from houses. So, our group decided to call for an outright ban on fracking in the

city limits. When the ban passed in 2014, that was the most important outcome ... for a while.

7. *Effects*: The effects are long-term outcomes in terms of values, institutions, society, and the natural world. We might also think of them
as the cumulative growth, decline, and shifting of values. Effects can
include new practices, new institutions, new laws or even entire legal
regimes, and new ecological realities like a restored or decimated
ecosystem.

In 2015, the Texas state legislature passed HB 40, a law that
ushered in a new balance of power between cities and the state when
it comes to fracking. HB 40 prohibits cities from banning fracking
and greatly curtails their powers for regulating it in any significant
way. This new legal regime is a net loss for cities and a net gain for
the state.

Using the social process tool in this case helped me to gain a clear analysis of the situation. From this, I was able to see what I thought the best policies would be. In other words, it helped me not only to see how the conflict
took shape as a paradox (i.e., community safety or state wealth). More than
that, it guided my ethical reasoning toward what I judged to be the right
resolution of the paradox for the common interest.

Molten Politics: Thinking Big

The "Overton Window" describes the range of policies and ideas that is
politically acceptable to the mainstream at any given time. We talked about
this above as the mainstream discourse. One way to think about effects
(from the social process) is to see them as shifts in the Overton Window.
For example, in the US, the legalization of same-sex marriage went from an
implausible scenario to a reality very quickly. These kinds of seismic shifts
are crucial for climate politics. To think about them, let's return to climate
change as a hyperobject.

As a hyperobject, climate change is molten: it is so massive that it warps
the fabric of spacetime, re-shaping natural and social systems, melting our
ethical and political categories (see Morton 2013). Ocean chemistry, the
Amazon rainforest, glaciers, economic supply chains, international laws,

geopolitics, and more are changing. The effects of climate change are rearranging things that were once thought to be set in stone. Perhaps, as a result, climate politics will upend our base values such that we reconceive the meaning of a good life and a just social structure.

So, climate politics requires us to think about the long-term, fundamental effects that interact with the more surface-level, shorter-term discourses and institutions. We need to not just see how the pieces move on the board (the social process), but how the board itself is melting, its lines and contours rearranging. For example:

- Could the "iron law" of climate politics (economic growth is untouchable) give way to some kind of post-growth, post-materialist, or steady-state global economy?
- Might nation states lose their integrity, giving way to a more chaotic mix of private–public security forces fighting along shifting borders and dwindling sources of freshwater?
- Will financial systems shift to a global currency pegged to the value of carbon? Will it become profitable to actually put carbon back into the soil and rocks?
- Will petrostates lose power in the wake of ascending "electrostates" like Chile, China, and Bolivia that contain the rare Earth minerals needed for an electrified world?
- Will mass migration (within and between existing national borders) force people to reconsider the privileges and obligations? Could it lead to more humane legal regimes for refugees? Or might it lead to more violence?
- Will bio-regional forms of sustenance re-emerge as global supply chains collapse in conditions of strife and scarcity? Will National Parks and wilderness areas be returned to native peoples?
- What about a pill that makes people stop craving meat? Could we imagine a future where people use drugs or even gene therapy to control desire or enhance empathy as climate policy strategies?
- Will it become normal for people to farm and eat insects as a source of protein that is more sustainable than beef, chicken, and pork?

Ok, we can't possibly imagine let alone prepare for all possible effects. Yet we tend to have the opposite problem, not imagining much past our current reality at all. We might attribute this to "system justification" (Jost

2020). One thing that the human or political sciences teach us is that people tend to crave safety, stability, and predictability. We are inclined, then, to defend and justify the status quo rather than upset the applecart. When we are born, we are thrown into a world of customs, conventions, habits, institutions, and laws. The tendency is to treat them as natural and normal: this is the way things are and always have been. This is true even for many people who are oppressed or exploited by the existing systems.

Every institution, law, road, pipeline, building, and scrap of infrastructure exists as the result of politics—that contingent process of humans sharing and shaping values. Yet once they are built, such things tend to be treated as given and beyond the realm of political change. That need not be the case. Indeed, to get to a thriving future, we'll need to accept that the world is molten. It is being shaped and re-shaped, and we need to learn how to play our parts—big and small—in the shaping.

This is one reason why it is vital to include Indigenous, multi-species, and multi-cultural perspectives in climate politics. Human history seen through diverse lenses shows us how many cultural constellations are possible, and it gives us alternative perspectives for seeing past our own tendencies of system justification. Things can be otherwise. It starts by asking whether the status quo structures really are justified. For that task, it will be helpful to pick up some tools from climate ethics.

Conclusion

Climate literacy requires critical thinking about politics as group decision-making activities that are neither as simple as applying science nor as irrational as the brute force of "might makes right." In this chapter, we picked up three tools to aid understanding of climate politics. These tools work on different levels of political analysis: a) discourse (speech), b) current institutions and actors, and c) the long-term, foundational shifts in both discourse and institutions. Just like the climate itself, climate politics is all about changes on various scales and timeframes.

On the level of discourse, we examined political reasoning as the making and resolving of paradoxes. This refines our understanding of how problems are framed and contested. On the level of political actors, we used the social process to organize our thinking about who is involved, what they want, the power and resources they possess, and how they pursue their goals. This puts discourse into the real-life contexts of social institutions. Finally, on the level of long-term, fundamental effects, we returned to climate change as a hyperobject that is not just debated through discourses (paradoxical reasoning) and that is not just moving pieces on the board (the social process). Climate change is also melting the board in ways that redraw categories and lead to revaluations of values.

Activities and Questions

1. Some climate scientists feel like their role in the linear model—conduct research and publish findings—is too limited and ineffectual. Should scientists be more directly involved in climate politics—perhaps engaging in protests, direct action, and other forms of activism? Research the words and deeds of the climate scientist James Hansen in this regard. Do you agree with him and other scientists who have taken political action?

2. Identify a controversial topic related to climate change and analyze the discourses involved in terms of paradoxes. How would you go about deciding which way this paradox should be resolved, that is, who has the stronger arguments and why?

3. Practice using the social process to analyze a climate-related topic. What insights does this yield? Does it give you a working map of the context? Does it miss important pieces? Does it offer guidance for your own involvement—that is, can you spot areas where more research is needed, common interests could be found, stakeholders are missing, etc.?

4. Stretch your imagination: what major effects will climate change bring? Do dystopian futures more readily come to your mind? What about a utopia of greater equity, biodiversity, and human flourishing? What categories need to melt and how does the Overton Window need to shift (if at all) in order to get us there? What might change: capitalism, private property, our sense of self and kin, etc.?

5. How can you best take part in climate politics? What passions and skills do you have? Do you want to work locally, nationally, or internationally?

Research organizations that inspire you and reach out to them to find out how you can get involved.

References

Arendt, Hannah. 1958. *The Human Condition*. Chicago: University of Chicago Press.

Briggle, Adam. 2015. *A Field Philosopher's Guide to Fracking*. New York: Liveright.

Clark, Susan. 2002. *The Policy Process: A Practical Guide for Natural Resource Professionals*. New Haven, CT: Yale University Press.

Dryzek, John. 2005. *The Politics of the Earth: Environmental Discourses*. Oxford: Oxford University Press.

Jost, John. 2020. *A Theory of System Justification*. Cambridge, MA: Harvard University Press.

Lasswell, Harold. 1936. *Politics: Who Gets What, When, How*. New York: Whittlesey House.

Morton, Timothy. 2013. *Hyperobjects: Philosophy and Ecology After the End of the World*. Minneapolis: University of Minnesota Press.

Olewe, Dikens. 2022. "CoP 27: Uganda-Tanzania Oil Pipeline Sparks Climate Row." BBC, October 24. https://www.bbc.com/news/world-africa-63212991.

Oreskes, Naomi, and Erik Conway. 2010. *Merchants of Doubt: How a Handful of Scientists Obscured the Truth on Issues from Tobacco Smoke to Global Warming*. New York: Bloomsbury.

Rauch, Jonathan. 2021. *The Constitution of Knowledge: A Defense of Truth*. Washington, DC: Brookings Institution Press.

Sarewitz, Daniel. 1996. *Frontiers of Illusion: Science, Technology, and the Politics of Progress*. Philadelphia: Temple University Press.

Stone, Deborah. 1998. *Policy Paradox: The Art of Political Decision Making*. New York: W.W. Norton.

CHAPTER 9

Climate Ethics and Justice

CO$_2$ Emissions Per Capita vs. GDP

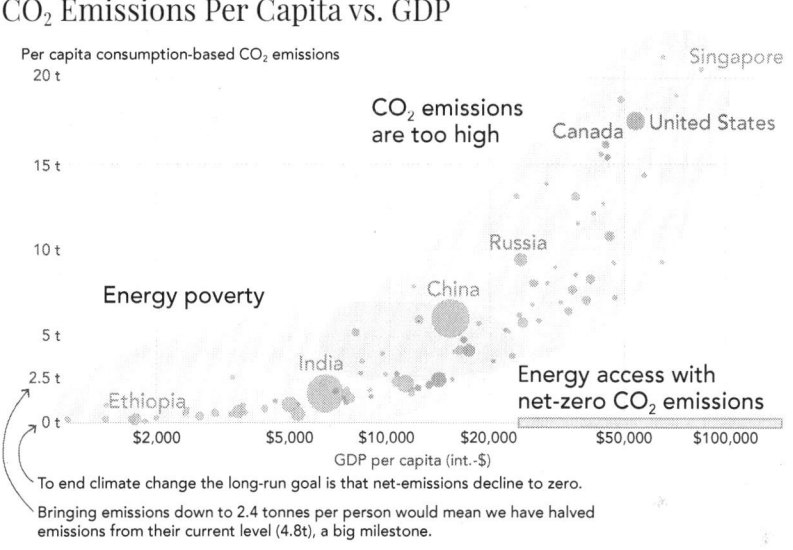

Per capita consumption-based CO$_2$ emissions

CO$_2$ emissions are too high

Energy poverty

Energy access with net-zero CO$_2$ emissions

Singapore

Canada   United States

Russia

China

India

Ethiopia

20 t

15 t

10 t

5 t

2.5 t

0 t

$2,000 $5,000 $10,000 $20,000 $50,000 $100,000

GDP per capita (int.-$)

To end climate change the long-run goal is that net-emissions decline to zero.

Bringing emissions down to 2.4 tonnes per person would mean we have halved emissions from their current level (4.8t), a big milestone.

FIGURE 9.1

Climate change belongs to what Aristotle called the practical sciences of ethics and politics, because our aim is not knowledge for its own sake, but action. The fundamental question for ethics is *how should we live*? In our daily, unthinking habits, we give answers to that question. But maybe those habits are wrong. Perhaps we can and should do better. The only way to find out is by *thinking*, that is, engaging in ethical inquiry.

Climate ethics in particular calls on us to think about how we should live, rather than just thoughtlessly carrying on with business as usual. After all, we participate in systems that contribute to the problems we have been analyzing with the help of the natural sciences. The question of how we should live is always related to our moment in history. How should we live in the Anthropocene or the Great Acceleration? How should we live in light of what we know about our impacts on the climate system? Of course, remember that "we" needs to be unpacked: who, exactly, should do what?

So, we've been engaging with climate ethics throughout the book. By turning now to ethics as a field of study, we can improve our skills by picking up more tools in our quest for climate literacy. Most importantly, we'll work on ethical reasoning, which entails not just saying "this is wrong (or right)," but "this is wrong (or right), because...." What reasons can we give in the search for answers to our normative questions about climate change?

In this chapter, we'll first use a story of climate activism to illustrate the fundamentals of ethical inquiry and action. Next, we'll survey the main normative ethical theories as traditionally understood. Then we'll turn toward environmental ethics, which ask if we need radically new thinking that decenters humanity as the sole site of intrinsic value. We'll conclude with a survey of climate justice, a concept that Figure 9.1 captures with a basic picture of some having too much and others not enough.

The Valve Turners

On December 7, 2015 a group of Canadian activists turned the emergency shut-off valve on an Enbridge pipeline carrying crude oil from western Canada to refineries in the east. In so doing, they called to the foreground of our thinking one of those hidden background sites of our commodious pattern of development. After stopping the flow of oil, these "Valve Turners" then locked themselves to the valve with U-locks around their necks. The next year, activists in the US did the same thing at five pipelines carrying oil from Canadian tar sands to Texas.

This is "direct action," because the activists used their power to achieve a desired end directly, without an appeal to others, such as voters through an election or authorities through a legislative process. The Valve Turners were arrested, because tampering with oil pipelines is illegal. In court, they used the necessity defense, arguing that breaking the law is justified given the threats posed by climate change.

In trial, their attorneys portrayed their actions as heroic efforts to prevent the existential risks of climate change. Attorneys for the pipeline companies, by contrast, called the Valve Turners extremists and likened them to terrorists. Both sides could agree on a factual description of what the Valve Turners did (indeed, the activists filmed their actions). That kind of knowledge wasn't at issue. Rather, the debate was about ethical knowledge. Was the Valve Turners' reasoning sound? Did they do the right thing? Was their action good or bad?

At the heart of this case is the key distinction between law and ethics. Just because something (like extracting and transporting oil) is legal does not make it ethical. More generally, just because people *do* act in a certain way doesn't mean they *should* act that way. Martin Luther King, Jr., who was also arrested for taking direct action, wrote in his 1963 "Letter from Birmingham Jail":

> There are two types of laws: just and unjust. I would be the first to advocate obeying just laws. One has not only a legal but a moral responsibility to obey just laws. Conversely, one has a moral responsibility to disobey unjust laws.

The ethical knowledge we seek hinges on this distinction: how can we tell a just law from an unjust law?

The Valve Turners argued that it is better to break the law than to obey it, because there is no time for incremental change. As justification for their actions, some of them cited an op-ed by climate scientist James Hansen (2012). In that article, Hansen wrote, "we are indeed in an emergency situation.... If fossil fuel emissions are not systematically and rapidly abated, then young persons and future generations will confront what can reasonably only be described as, at best, an inhospitable future." In 2013, Hansen was arrested for participating in protests of oil pipelines.

In the Valve Turners case, a Montana judge didn't buy the argument about imminent harm and refused to frame the case as a matter of climate change, stating: "The energy policy of the United States is not on trial" (Brown 2018). Indeed, legislatures in Alberta and Texas reacted to protests at oil and gas sites by passing "critical infrastructure" laws that increase the penalties for interrupting the operations of pipelines and other infrastructure deemed essential to security.

Recall our "what is it?" questions. What is an oil pipeline: a climate risk or critical infrastructure? What is shutting down a pipeline: heroism or terrorism? Here we see how the paradoxes of politics are also central to ethics. They are a matter of *descriptive ethics* or how a situation gets framed. What moral language is used and how are important terms like "security" defined? Oil pipelines are *both* critical infrastructure for security *and* threats to security. By now, we should be used to thinking in such tangles. Let's turn to ethical theory for some more tools to help us out.

Ethical Theory

Normative ethical theories guide our thinking on our quest toward ethical knowledge. They do so by offering different explanations for what makes something right or wrong, good or bad. In this way, they help us distinguish just from unjust laws. They do so by supplying the vocabulary that follows the "because" clause in ethical reasoning: This is wrong (or right), because {X ethical theory tells us so for these reasons}.

In this section, we'll look at three such theories. First, though, let me note that I think of Aristotle's virtue ethics as a kind of meta-theory. By that, I mean that part of being a good person is having the skills (virtues) to reason and act wisely in different contexts. A virtuous person has courage and good judgment to know how to use the right theories in the right ways for the right purposes. Ethical theories are useful heuristics for action but, like the problem orientation and the social process, they don't offer easy recipes for hard decisions.

You can keep the Valve Turners in mind as we survey these theories. Let's also add another case: meat-eating. Figure 9.2 shows the GHG emissions from different food products. Clearly, meat contributes the most emissions. So, is it unethical to eat meat?

As with most sources of emissions, meat-eating is linked to human well-being. Meat is a vital source of nutrition, yet it is also not a dietary necessity for most people. This is not necessarily an all-or-nothing question. Maybe the trick is to find the right amount or type of meat consumption. For example, chicken clearly creates less emissions than beef. And there is a wide variety of consumption behaviors: Indians consume less than 5 kilograms (kg) of meat per year, Brits consume about 80 kg, and Americans consume nearly 130 kg. With these complexities in mind, let's survey the three theories.

Consequentialism

According to this theory, an act is right or wrong according to its consequences. The right action is the one that produces the most good. Of course,

Greenhouse Gas Emissions Per Kilogram of Food Product

Emissions are measured in carbon dioxide-equivalents.[1] This means non-CO_2 gases are weighted by the amount of warming they cause over a 100-year timescale.

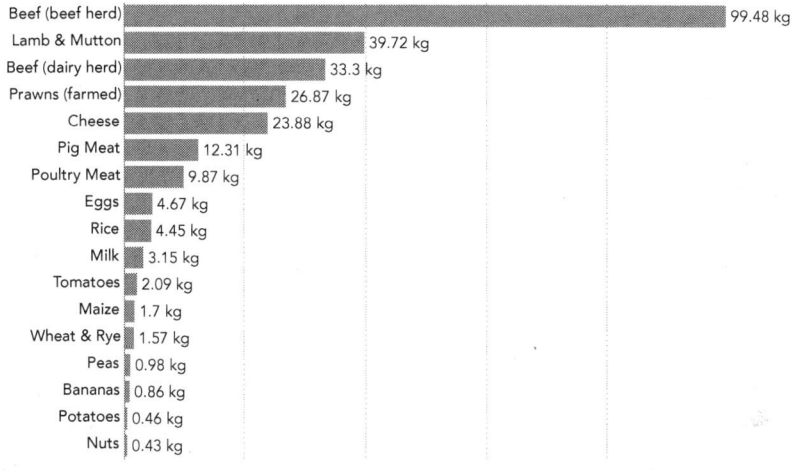

Beef (beef herd) — 99.48 kg
Lamb & Mutton — 39.72 kg
Beef (dairy herd) — 33.3 kg
Prawns (farmed) — 26.87 kg
Cheese — 23.88 kg
Pig Meat — 12.31 kg
Poultry Meat — 9.87 kg
Eggs — 4.67 kg
Rice — 4.45 kg
Milk — 3.15 kg
Tomatoes — 2.09 kg
Maize — 1.7 kg
Wheat & Rye — 1.57 kg
Peas — 0.98 kg
Bananas — 0.86 kg
Potatoes — 0.46 kg
Nuts — 0.43 kg

1. **Carbon dioxide-equivalents (CO_2eq):** Carbon dioxide is the most important greenhouse gas, but not the only one. To capture all greenhouse gas emissions, researchers express them in 'carbon dioxide-equivalents' (CO_2eq). This takes all greenhouse gases into account, not just CO_2. To express all greenhouse gases in carbon dioxide-equivalents (CO_2eq), each one is weighted by its global warming potential (GWP) value. GWP measures the amount of warming a gas creates compared to CO_2. CO_2 is given a GWP value of one. If a gas had a GWP of 10 then one kilogram of that gas would generate ten times the warming effect as one kilogram of CO_2. Carbon dioxide-equivalents are calculated for each gas by multiplying the mass of emissions of a specific greenhouse gas by its GWP factor. This warming can be stated over different timescales. To calculate CO_2eq over 100 years, we'd multiply each gas by its GWP over a 100-year timescale (GWP100). Total greenhouse gas emissions – measured in CO_2eq – are then calculated by summing each gas' CO_2eq value.

FIGURE 9.2

as we saw in our discussion of uncertainty, we often cannot predict consequences. So, this needs to be modified: An act is right if it is reasonably expected to produce the greatest balance of good or the least balance of harm. This raises the key question of how much foresight (thinking about potential consequences) should be done prior to action. Recall that climate change is largely the *unintended consequence* of the modern, fossil-fueled development project. How much can and should we expand our intentionality, our forethought, prior to action?

Utilitarianism is the most popular version of this theory, where the "good" is understood as "utility." As we'll see in the next chapter, utilitarianism is the ethical theory behind much of policymaking, especially in the form of cost–benefit assessments and the precautionary principle.

Key questions for utilitarianism include:

a. Who gets counted in the ethical calculation of utility? How far and wide do we track consequences? For example, should we consider future generations?
b. What is meant by "utility"? Is this happiness, well-being, pleasure, the absence of pain?
c. How are different utilities measured and compared?
d. How sure must we be about the likely consequences of our actions before we can act?

The Valve Turners made a consequentialist argument that their actions would bring about the greatest good or least balance of harm. Clearly, the pipeline companies disagreed, arguing that disrupted oil supplies are a much greater and more imminent harm than climate change.

In the case of meat-eating, the climate argument is one form of consequentialism: meat-eating contributes to climate harms. Peter Singer (1975) makes the utilitarian case for vegetarian diets and animal welfare in a different register. He argues that climate dangers to *human* welfare are the wrong reason to oppose meat-eating. Even if meat was climate neutral, we should still not eat it, because animals have the capacity to feel pleasure and pain. This means that they should count in our calculations. Meat eating is speciesism (like racism or sexism) where we treat human interests above those of animals for no ethically significant reason. Animals have moral status—they deserve ethical consideration—because they can feel pleasure and pain. Surely the pleasure we derive from eating beef does not outweigh the pain, suffering, and death of a cow (a sentient creature), especially when we can make other choices.

Deontology

This ethical theory derives its name from a Greek word meaning "duty"—an unconditional obligation. For a deontologist, what makes an act right or wrong is not the consequences that result from it, but rather the motive that gives rise to the action. The ethical life is one where we act *for the right reasons* (no matter what the consequences!).

Consider how a consequentialist theory might lead us to some troubling conclusions. What if your great Aunt is a mean person who has lots of money. Couldn't you reason that killing her in order to inherit her wealth would be for the best? After all, everyone hates her and you could give all her money to worthy charities for youth. So, the ethical calculus looks pretty good: one elderly life sacrificed to benefit thousands of children. If this sounds wrong to you, deontologists would say that is because your great Aunt has rights, especially the right to life. Such rights language is harder to capture (though not impossible) in a consequentialist theory.

The Valve Turners appealed to the rights of future generations to be left a planet that is not "inhospitable." The pipeline companies argued on the basis of their private property rights as protected by legal contracts. The activists might reply by questioning the ethics of thinking of the land or resources in terms of private property at all.

In the case of meat-eating, Tom Regan (1983) argues that the status quo is wrong not for climate reasons and not because of animal pain and suffering. After all, we could lower emissions from meat and we could keep animals sedated in cages such that they feel no pain. No, he argues, the reason to stop eating meat is because the current system treats animals as our resources without any intrinsic value or rights of their own. Using animals for food is wrong, because they are "experiencing subjects of a life" with their own volition. They have life plans that deserve our respect. We cannot justify the use and confinement of some (even if it is painless) for the pleasure of others. Yet, could one counter that people have a right to make their own dietary choices? What is a right, anyway? How do we know when they exist?[1] How do we weigh competing rights claims?

Relational Ethics

This is really a collection of theories that sometimes goes by the name "ethics of care" or "feminist ethics." From this perspective, the other theories are backwards. They both try to find an impartial view and distill some essence that belongs to individuals and derive moral status from this. Animals, for example, have the capacity to feel pleasure and pain (consequentialist) or are experiencing subjects of a life (deontology). This makes it sound like ethical values are things that exist prior to and independent of us, and that we discover them "out there."

But maybe this is not how our ethical lives work. What gives something value is not some property it possesses that we can discover when we take

---

1   Years ago, when the Canadian Rugby Team was forced to cancel a match against South Africa (because of a boycott against apartheid), members of the team objected that this violated their "right to play." Is this a right others hadn't noticed before?

an impartial view. Rather, we are enlivened with values by the relationships that constitute who we are: mothers, brothers, sons, colleagues, neighbors, friends, citizens, etc. These webs of relationships sustain and shape us in ways that are intrinsic to deciding how we should be treated. In other words, we create and confer values on others. We don't discover values as pre-existing things; we invent them! We know that humans are valuable, because we give each other names and celebrate birthdays, graduations, weddings, etc. We are valuable, because we are valued in caring relationships.

Cora Diamond (1978) argues from this perspective that if we want people to stop eating meat, we need to form stronger relationships with animals. We need to *humanize* them in this way—treating them as "fellow creatures." After all, we don't eat our pets, because they have names and share our home—pets just aren't constituted by relationships as "things we eat." Ethics requires us to get relationships right (but what does this mean?) and act appropriately within those bonds.

Relational-ethics theorists most often talk about the ethical commitments that arise from our relationships with family, friends, lovers, and nearby community. But in the Valve Turners case, the pipeline forms an ambiguous network of relationships that are legal, political, economic, and ecological. The competing sides highlight different aspects of these relationships to make their case. What's most essential about climate ethics here is the idea that perhaps we need to radically reconsider our relationships in

Environmental Ethics

	CONSEQUENTIALISM	DEONTOLOGY	RELATIONAL ETHICS
WHAT MAKES AN ACT RIGHT/ WRONG, GOOD/ BAD?	The consequences of the act	The motive that initiates the act	The suitability for the relationship at hand
VALVE TURNER CASE STUDY	Climate change as imminent threat? Consequences of disrupted oil flows?	Rights of future generations? Property rights?	Pipeline as a web of relationships ... a new ethic for humanity in relation with Earth?
MEAT-EATING CASE STUDY	Singer: wrong, because animals feel pain	Regan: wrong, because animals are experiencing subjects of a life	Diamond: wrong, because animals are fellow creatures

FIGURE 9.3

light of the enormous powers our fossil-fueled technologies have given us. Maybe behind the Valve Turners' actions is the idea that humanity is acting hubristically and irresponsibly in relation to the rest of the biosphere. We need to learn how to form an ethical relationship with the Earth as a living community. This leads us to environmental ethics as a field of inquiry that questions whether these traditional ethical theories are adequate for the task of living well in the Anthropocene.

Beyond Anthropocentrism: Environmental Ethics

The field of environmental ethics started to take shape in the latter half of the twentieth century. I see climate ethics as an outgrowth of this development. The core idea is that times have changed and, if we are to survive, we need a radically new ethic. Modern Western ethics—the value system that has given rise to the Anthropocene—tends to be anthropocentric. According to this predominant ethic, only humans possess intrinsic value (the term means human-centeredness). All other species as well as ecosystems are only instrumentally valuable as resources for our use and enjoyment.

In terms of our theories above, this means that it has largely been assumed that:

a. Consequentialism: only consequences for humans matter and only human utility (goods/harms, pleasures/pains) is taken into account.
b. Deontology: only human rights and freedoms matter in forming duties, responsibilities, and contracts.
c. Relational ethics: only human relationships matter when considering the ethical life and fitting conduct toward one another.

Of course, thinkers like Singer, Regan, and Diamond have pushed back against this orthodox view in their own ways (often disagreeing with each other!). Yet as trends in meat-eating, habitat loss, GHG emissions, and species decline indicate, anthropocentrism is still in full force.

Environmental ethicists have been arguing that we can think of progress in ethics as expanding a circle of who and what counts, that is, who and what are ethically considerable. Traditional ethics have urged us to move from egocentrism (only counting one's self) to tribal-centrism (only counting one's group) to anthropocentrism (accounting for the intrinsic value and dignity of all humans). Now, the argument goes, our enormous powers create an ethical imperative to expand to biocentrism (recognizing the intrinsic value of all living things) or ecocentrism (recognizing the intrinsic value of entire wholes like ecosystems or even Gaia,[2] planet Earth). Increasingly, ethicists are turning to Indigenous and non-western traditions to help in articulating these different perspectives and expanded relationships.

Progress in Ethics

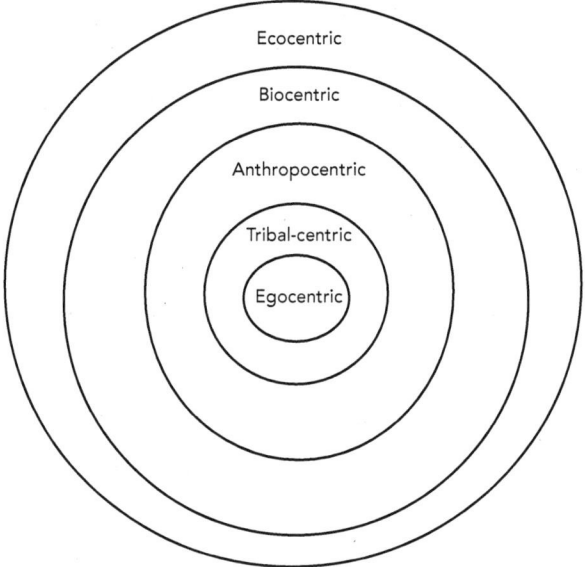

FIGURE 9.4

There is a divergence in climate ethics between a "deep" and a "shallow" approach. As we've seen, the predominant approach is shallow in that it does not question anthropocentrism. Many involved may not even think to question this deeply held assumption. Others do, but conclude it is simply too radical and unworkable: we just won't see such a massive "revaluation of

---

2   In Ancient Greek mythology, Gaia was the mother of the other gods and of all life on Earth. Some eco-spiritualist movements have taken her up as symbol of the Earth as a holistic environment. And some scientists have proposed theories that view Earth as akin to a living organism—as a self-regulating and synergistic system that helps to maintain conditions suitable for life. This is known as the Gaia hypothesis.

values," so climate solutions need to be grounded in our current anthropo-centric ethics. For example, we should frame things as a matter of enlightened self-interests for humans. So, rather than talk about the intrinsic value of a wetland we can try to calculate the "ecosystem services" (like water filtration and storm surge protection) that it provides for people.

And yet, as with last chapter, I think it is important to consider more radical approaches to climate change. Our ways of thinking about ethics just may not be up to the task of handling the new reality that we are creating. For example, Stephen Gardiner (2010) calls climate change "a perfect moral storm," because it brings together three factors that challenge our ability to act ethically:

1. Dispersion of causes and effects: GHG emissions spread far in space and time from their source, making it exceedingly difficult to track consequences and pin down responsibility.
2. Fragmentation of agency: as a collective action problem (Chapter 3) climate change invites actors to cheat or free-ride in their own self-interest, even though they recognize that cooperation would be best for the whole.
3. Institutional inadequacy: solving collective action problems requires enforceable rules, but the current international political system struggles to hold actors accountable for their behaviors, including nation states and multi-national corporations.

Gardiner argues that we are ill-equipped to deal with the ethical questions posed by climate change, because they involve thinking on temporal and spatial scales that far exceed the assumptions behind all previous ethical theories and existing institutions. Others have made similar arguments that we need a deeper ethic, one that changes our fundamental worldview about humanity and our place in nature. Let's look briefly at two related ways to argue for a radically new ethic.

New Technologies Bring New Responsibilities

Hans Jonas (1974) was among the first to argue that modern technology has "changed the nature of human action" in a way that calls for "a change in ethics as well." All previous ethics, he argued, could assume that humans were powerless to impact nature on any meaningful scale. Ethics was thus limited to human–human interactions, "All traditional ethics is *anthropo-centric.*" Ethical consideration was bounded by horizons that were close in time (the span of a life) and near in space (neighbor with neighbor, friend with foe, superior to subordinate). It follows that the knowledge required to act well was also limited and available to anyone of good will.

All of this has changed due to modern fossil-fueled technological powers. Nature is now vulnerable to our actions, which have much wider horizons. It follows that the knowledge needed for ethics outstrips our own experience—this is why we require science to inform our deeds that have far-flung, hidden consequences. Knowledge, Jonas argues, has become a duty, and that knowledge must be commensurate with the scale of our action. This is all unprecedented and requires a "new concept of duties and rights." Jonas argues that the biosphere, now subject to our power, has become our trust and makes a moral claim on us to care for it not just for our own enlightened self-interest but also for its own sake.

Jonas was unsure about how we could bring about this new ethic to match our new powers. He suggested that we need to cultivate a new kind of fear. Instinctive fear happens automatically. When we hear a noise at night in the dark, for example, our heart races and we go into alert mode. Yet climate change and other environmental problems are "slow disasters" or "slow-onset events" that do not spark the same instinctive reaction. So, Jonas argued that our task—through art, media, politics, and education—is to cultivate a higher-order fear so that we are motivated to act appropriately given our new relationship with the planet.

Ingmar Persson and Julian Savulescu (2012) have a much different idea for altering human ethical behavior. They similarly argue that for 99 percent of our history, humans evolved and thrived under vastly different conditions. Our interactions were mostly local with relatively immediate consequences. This means that our brains are wired to think in these terms (see Duhaime 2022). So, Persson and Savulescu argue, maybe we should genetically engineer humans with the cognitive and emotional abilities to have the expanded thinking and empathy required now by our technological powers. If education doesn't work, maybe this is the only way to make us fit for the future that we are creating. We can think of this also in terms of adaptation: we just need to accelerate our own evolutionary process of adapting to a rapidly changing environment.

New Sciences Bring a New Image of Humanity

Aldo Leopold (1949) was among the first to think about ethics in light of the relatively new science of ecology. He argued that an ethic understood ecologically is a limitation on the freedom to act in the struggle for survival. Ethical limits have their origins in the tendency of interdependent organisms to evolve modes of cooperation like symbioses. Leopold thought that the story of human ethics was one of extending circles of ethics from individuals to more and more members of human society as we become entangled in more webs of interdependence. It is now an ecological necessity, he argued, for humans to extend ethical consideration to the land (soils, waters, plants, and animals). We need an "ecological conscience."

Leopold was one of the first to develop an *ecocentric* ethics—a form of ethical holism that locates intrinsic value not just in humans or even other individual animals or organisms, but rather in ecosystems as a whole. Arne Naess (1973) and others would later call this "deep ecology." No longer can we just treat the land as only instrumentally or economically valuable, that is, as matter of serving only our self-interests (even if enlightened). We can't just look at a forest and see lumber, a prairie and see grazing land for cattle, or a mountain and see mining locations. Evolution has taught us that we are kin with all living things, and ecology has taught us that we participate in the same webwork flows of materials and energy. So, we need to evolve an ethic of right relations with our fellow members of the living community.

Leopold summed up his view of an extended relational ethic like this, "a land ethic changes the role of *Homo sapiens* from conqueror of the land-community to plain member and citizen of it. It implies respect for his fellow-members, and also respect for the community as such" (204). So, the "what is it?" question here is about "who are we?"—are we conquerors or plain members of the biotic community? Leopold offered a new ethical theory to guide our actions as plain members:

A thing is right when it tends to preserve the integrity, stability, and beauty of the biotic community. It is wrong when it tends otherwise. (224–25)

Leopold set the agenda for much of environmental and climate ethics that has followed. It is a two-fold strategy. First, criticize the predominant anthropocentric, mechanistic worldview that has created the climate/ecological crisis (recall science as a worldview from Chapter 5). Second, help to give birth to a new holistic, ecocentric worldview that would put an end to the crisis (a new science gives birth to a new worldview). Let's conclude by thinking about climate justice in light of these developments in ethical theory.

Climate Justice

As with other aspects of ethics, we've already been thinking about climate justice, especially the glaring global inequities: the wealthiest 1 percent are responsible for 15 percent of GHG emissions and the wealthiest 10 percent emits about half of GHGs. Meanwhile, the poorest half of humanity is responsible for just 7 percent of emissions and is most vulnerable to climate hazards. Figure 9.1 offers one way to portray this in a chart. For an example of a story, consider the Marshall Islands, which are being submerged by rising seas, yet the Marshallese make a negligible contribution to climate change. On a per capita basis, their GHG emissions are less than half the global average. Saving the Marshall Islands will take "radical adaptation" measures that will cost $1 billion. Should this happen and who should pay for it?

The philosopher Henry Shue (2014) calls this the central question of climate justice: "[h]ow can we limit the dangers resulting from climate change without driving additional hundreds of millions of people into poverty?" (4). For Shue, the relationship between rich and poor nations is the most important focus for climate ethics. The rich have exploited carbon resources to achieve high levels of economic development, whereas the poor have not developed and must do so now in the context of new vulnerabilities on a warming planet and international pressures to limit their carbon resource development (recall the EACOP pipeline example in Africa from Chapter 8). For Shue, one key element of climate justice is guaranteeing that poor countries have the right to levels of energy (but not necessarily emissions) needed for their development.

In Chapter 3, we saw that the main goals of safety and well-being should be pursued in ways that "reflect equity and the principle of common but differentiated responsibilities and respective capabilities" (from Article 2 of the Paris Agreement). In other words, as with so much involving climate change, justice really isn't a separate issue at all. Questions of justice are part of the tangle. Indeed, they are built into the two main goals. To talk of safety and well-being is to imply questions about who deserves what levels of safety and well-being and who has what obligations—financial or otherwise—to make that happen.

For example, the philosopher Walter Sinnott-Armstrong (2010) considers climate change as a collective action problem to argue that there is group level responsibility to act but no individual level responsibility. No individual is morally responsible for climate change and no individual's acts would matter for reaching climate goals. Sinnott-Armstrong thinks it would be absurd to task individuals with responsibility, because even a slight action (like taking a jog around the block) emits excess CO_2. If we had a moral duty to refrain from creating excess GHG emissions, we would be obliged to lie as motionless as possible. Thus, the absurdity. His conclusion is that the governments of developed nations have a duty to address this collective action problem.

In another example of how climate justice cuts across the tangles of climate policy, one study pictured the atmosphere as a global commons and asked what a fair share for each country would be with respect to the 1.5°C and 2°C targets (Fanning and Hickel 2023). What they call the Global North is on track to overshoot its fair share by a factor of three (see Figure 9.5). Calling this unjust, the authors then placed a monetary value on the excess emissions (social cost of carbon) and distributed the money to low-emitting countries as a form of compensation based on how much of their fair shares would be appropriated by over-emitting countries. They conclude that the Global North owes the Global South a staggering $170 trillion in what might be called climate reparations. The United States would owe the most in this scenario at $80 trillion, which works out to $500,000 for every taxpayer. Do they really owe that much? What is climate justice?

Fair Shares of Global Carbon Budgets

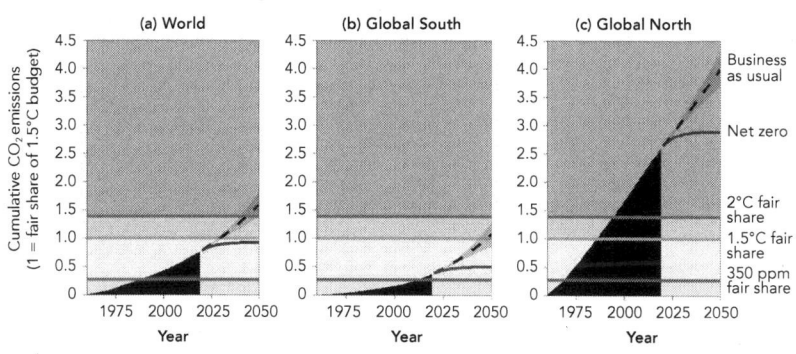

FIGURE 9.5

Justice is central to ethics, law, and politics, and its meaning is often contested in disagreements about the rational justification for acting one way rather than another (see MacIntyre 1988). One way to think through these disagreements is to consider four types of climate justice.

Climate Justice as Fair Distribution

Distributional justice is a matter of giving to each what they are due. It means a fair distribution of the benefits and costs related to climate change. The environmental justice scholar Kristin Shrader-Frechette (2002) put it this way: "a fair or equitable distribution of society's technological and environmental risks and impacts" (24). This view tends to picture justice primarily as an *outcome*—what matters is who gets what. Fair outcomes might be distributive schemes where everyone gets what they deserve or need (who determines that?), or what they are happy with, or what everyone else has (total equality). What are just climate outcomes? Is it a world where:

* everyone has the same level of emissions?
* all people are equally protected from climate hazards?
* the best people get the most rights to produce emissions?
* everyone has their basic needs met? Or everyone leads a modern developed lifestyle?
* all species flourish?
* everyone pays their fair share for climate damages?

The first formulation is hotly debated in the climate ethics literature (see Jamison 2013). Perhaps fairness means that each person is allocated the same number of GHG emissions permits. Is this a reasonable proposal, and how would it be administered and monitored? The last formulation (paying fair shares) tends to be the most common and was the impetus for a 2009 global agreement at COP15 of the UNFCCC in Copenhagen. There, developed nations pledged to contribute $100 billion annually for climate action in developing countries. Perhaps not surprisingly, contributions have fallen short of the pledge, though later COP treaties (like COP 28 in 2023) began to cover that shortfall.

Climate Justice as Participation

We can also think of justice as a *process* and not just an outcome. One formulation focuses on participation: Justice means the fair treatment and inclusion of all people (and non-humans?) in the making of decisions about climate change. Or we might say: everyone has a fair say in decisions that impact them. Central to this notion are the concepts of autonomy and informed consent—others can't just treat you instrumentally or discount you when their decisions affect your life. Leopold might point out that this is precisely how we have treated the prairies, forests, and waters.

It can be hard to know how to implement this view of justice. For example, should everyone potentially impacted by my emissions be involved every time I decide to drive my car? This notion of justice as process or democratic participation can also bump heads with outcome-oriented notions of justice. For example, the outcome of a swift energy transition to a net-zero economy will require lots of new renewable energy generation and transmission projects. Those projects often face resistance from people who want a fair say in how their homes, neighborhoods, or property rights will be impacted.

Another important question pertains to *scope*: to whom or what do principles of justice apply (who has an obligation to whom)? Does justice apply only to people who stand in a certain relationship with each other? For example, maybe justice only extends as far as national borders, because these determine who is obligated to pay taxes that might serve to redistribute wealth. Or maybe justice applies not just internationally but across species. *Multispecies justice* is the idea that the interests of non-humans should also count both in terms of process and outcome. One fascinating case here is the Atrato River, which is classified as a living entity with rights according to constitutional law in Colombia.

Further what about *intergenerational justice*? How should we account for the interests of future generations who cannot be present to participate in our decisions? Economists often talk about the "tyranny of the present," whereby future generations are wholly dependent on the current generation for the world that they will inherit. What might it look like to give future generations some voice or representation in making decisions that will impact them? We'll return to this issue in the next chapter with a discussion about discount rates in climate policymaking.

Climate Justice as Recognition

Another way to frame climate justice is in terms of a group's right to self-determination or autonomy. This is especially important for Indigenous and non-western communities that do not—and do not want to—fit into the pattern of "development" that we have been examining. Don't they have a right to their own conceptions of well-being?

Most visions of justice as distribution or participation take place *from within* the predominant pattern of development. For example, a community that is framed or defined as "poor" will be said to need a fair distribution of the "benefits of development." Or, a community will be seen as marginalized and in need of a voice to participate in the official institutions of decision-making. Yet, it may be that the community in question doesn't see themselves as poor or marginalized. If that is the case, then projects of "justice" start to look like colonialism: imposing a certain way of life onto others (supposedly for their own good). Is the development project that we've been pondering itself a form of justice (everyone deserves development) or an injustice (forced enrollment into an unwanted way of life)?

As the Indigenous studies scholar Glen Coulthard (2014) writes, "Communities are not fighting for the 'distribution of risks and impacts,' but for the right to live 'in relation to one another and the natural world in non-dominating and non-exploitative terms'" (13). Justice as recognition gets us back to our "what is it?" or "who are they?" questions. To *misrecognize* a group is to impose your standards and expectations on them, even if well-intentioned. True recognition means listening to other ways of being and allowing them to flourish.

Climate Justice as Retribution

Retributive justice pertains to efforts to find a proportionate punishment for acts of wrongdoing. This primarily arises in climate change through discourse about "loss and damage" or "climate reparations." As we've seen, attributing specific damages to climate change is difficult, but it is clear that climate change is already contributing to harms: destruction of homes and infrastructure, loss of crops, and deaths. Some of these losses are more difficult to quantify than others, such as the loss of local knowledge, territory, or mental well-being. Retributive climate justice efforts call on big emitters to compensate vulnerable groups for the losses and damages that they have incurred due to climate change.

Though the idea of climate reparations was long debated, it wasn't until COP 28 in 2023 that a fund was officially created to source money from developed nations to help pay for climate damages in developing ones. The Climate Impact and Response Fund (as it is known) will become an important site of climate politics. Developing nations will use it to lobby for more financing for their adaptation and mitigation efforts, and developed nations will wrestle over how much to contribute. When the fund was created, the US remained adamant that it has no legal obligation to contribute to the fund, and it insisted that it is not liable for damages in other countries. The US even balked at the term "loss and damage," because it implies moral culpability for harms. Yet, isn't that an accurate characterization of the situation?

Conclusion

How should we live? In this chapter, we picked up some tools to help us answer this question as it crops up in debates and decisions about climate change. We focused on ethical reasoning skills by considering the key distinction between just and unjust laws and by surveying three kinds of ethical theories. This led us to environmental ethics, which challenges the anthropocentrism and the narrow time and space scales of traditional ethical theories. We concluded by examining four ways to picture climate justice: distribution, participation, recognition, and retribution.

This chapter and the one before it discussed political and ethical theory in general terms. The last three chapters will hone in on climate policy—the contexts where political and ethical debates manifest in real world decisions. We'll first explore some fundamentals and frameworks for understanding climate policy. Then in the last two chapters we'll turn to some of the big areas of climate policymaking: mitigation, adaptation, finance, geoengineering (or climate intervention), and rewilding.

Activities and Questions

1. What do you think about the direct action of the Valve Turners? Are they justified in breaking the law—why or why not? Do you think this is an effective form of climate action?

2. Agriculture accounts for roughly 25 percent of human GHG emissions. The biggest reason for this is meat production and consumption. Does this mean that it is unethical to eat meat? Why or why not?

3. Do you think that climate reparations are a good idea? If so, what would a fair process and a fair outcome look like?

4. Do you agree with Sinnott-Armstrong that there are only governmental obligations and no individual-level obligations to act on climate change? If so, does that mean individuals can do whatever they please and not worry about climate change at all? Or would they have a duty to lobby their government to act?

References

Brown, Alleen. 2018. "Environmental Extremism or Necessary Response to Climate Emergency? Pipeline Shutdown Trials Pit Activists against the Oil Industry." The Intercept, March 21. https://theintercept.com/2018/03/21/pipeline-protest-necessity-defense-tar-sands/.

Coulthard, Glen Sean. 2014. *Red Skin, White Masks: Rejecting the Colonial Politics of Recognition.* Minneapolis: University of Minnesota Press.

Diamond, Cora. 1978. "Eating Meat and Eating People." *Philosophy* 53 (206): 465–79.

Duhaime, Ann-Christine. 2022. *Minding the Climate: How Neuroscience Can Help Solve Our Environmental Crisis.* Cambridge, MA: Harvard University Press.

Fanning, A.L., and J. Hickel. 2023. "Compensation for Atmospheric Appropriation." *Nature Sustainability* 6: 1077–86. https://doi.org/10.1038/s41893-023-01130-8.

Gardiner, Stephen. 2010. "A Perfect Moral Storm." In *Climate Ethics: Essential Readings,* ed. Stephen Gardiner, et al., 87–98. Oxford: Oxford University Press.

Hansen, James. 2012. "Game Over for the Climate." *New York Times,* May 9.

Jamieson, Dale. 2013. "Climate Change, Consequentialism, and the Road Ahead." *Chicago Journal of International Law* 13 (2): 439–68.

Jonas, Hans. 1974. *Philosophical Essays: From Ancient Creed to Technological Man.* New York: Atropos Press.

Leopold, Aldo. 1949. *A Sand County Almanac: And Sketches Here and There.* Oxford: Oxford University Press.

MacIntyre, Alisdair. 1988. *Whose Justice? Which Rationality?* Notre Dame, IN: University of Notre Dame Press.

Naess, Arne. 1973. "The Shallow and the Deep, Long-Range Ecology Movement. A Summary." *Inquiry* 16 (1–4): 95–100.

Persson, Ingmar, and Julian Savulescu. 2012. *Unfit for the Future: The Need for Moral Enhancement.* Oxford: Oxford University Press.

Regan, Tom. 1983. *The Case for Animal Rights.* Oakland, CA: University of California Press.

Shrader-Frechette, K. 2002. *Environmental Justice: Creating Equality, Reclaiming Democracy.* New York: Oxford University Press.

Shue, Henry. 2014. *Climate Justice: Vulnerability and Protection.* Oxford: Oxford University Press.

Singer, Peter. 1975. *Animal Liberation: A New Ethics for Our Treatment of Animals.* New York: HarperCollins.

Sinnott-Armstrong, Walter. 2010. "It's Not My Fault: Global Warming and Individual Moral Obligations." In *Climate Ethics: Essential Readings*, Stephen Gardiner, et al., 332–46. Oxford: Oxford University Press.

CHAPTER 10

Climate Policy Fundamentals and Frameworks

The Colorado River Basin

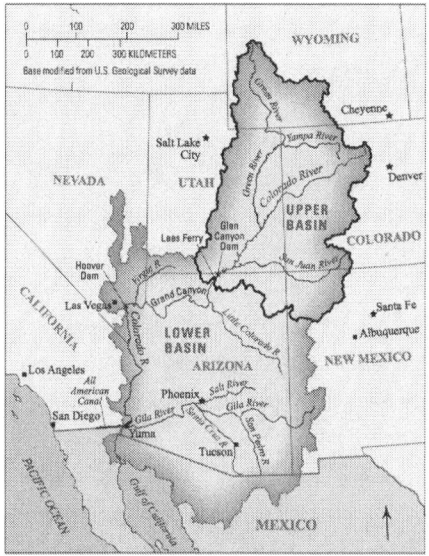

FIGURE 10.1

The Colorado River has always changed seasonally with the weather—surging with spring snow melts and subsiding in drier months. Yet there are also now climatic changes happening, that is, changes to the way things change. Decades of warming have been drying out the American West, causing less water to flow in the river. At some point a prolonged drought (weather) becomes aridification (climate)—a new climatic regime takes over. It may be that the American West and the Colorado River basin are undergoing such a shift.

At the same time, the population relying on the Colorado River for water, food, and power has skyrocketed. The American West went from 9 million residents in 1920 to over 78 million in 2020. The Colorado River has become a vital part of the US economy. It irrigates 15 percent of the nation's farmland, produces 90 percent of its winter vegetables, and supplies its two largest reservoirs—Lake Mead and Lake Powell.

The 1922 Colorado River Compact divided the river among two groups of states—the upper and lower basins. In deciding how much water each basin would get, however, the policymakers greatly overestimated the river's average flow. They also did not account for water rights claims by Mexico and many tribal states. As a result, more water was allocated than actually existed. Population growth, intensified agriculture, and the prolonged drought have added to this initial overallocation to precipitate a crisis. Simply put, people are going to have to use less water.

This is easier said than done. Cities, farmers, ranchers, tribes, environmental organizations, and others all have different ideas about what cuts should be made. Of the 1.9 trillion gallons of river water used annually, 12 percent is residential, 9 percent is commercial and industrial, and 79 percent is agricultural (Shao 2023). The vast majority of agriculture usage is for livestock feed (alfalfa, hay, grasses, and corn silage). In 2023, a temporary solution was announced whereby the US federal government agreed to pay $1.2 billion for various parties to voluntarily forego using a total of three million-acre feet of water. Experts estimate, though, that cuts four times that size will be needed to reach a sustainable level of water use and allow the reservoirs to recover (Jones 2023).

The Colorado River is governed by the physical laws of a changing climate. In terms of human control, it is managed by numerous compacts, laws, treaties, and guidelines collectively known as "the Law of the River." This is a good illustration of the complexities of climate policies. Such policies are not made like a statue as a product of single-minded design and execution. The Philosopher King might rule in this way and maybe climate policy will someday be "produced" by Artificial Intelligence. But for now, climate policies are not algorithms or recipes.

Rather, policies are the result of what the political theorist Deborah Stone (1988) called a "struggle over ideas." People are motivated by ideas that become the basis for shared meaning and for shared interests like farmers, ranchers, and conservation groups. What *is* the Colorado River and what is it *for*? It depends on who you ask: it is a source of water, electric power, recreation, irrigation, and habitat. Policymaking, Stone writes, is a "constant struggle over the criteria for classification, the boundaries of categories, and the definition of ideals that guide the way people behave." Policymaking is all about different ways to see reality and frame problems.

Focusing on policy is so vital for climate change, because as we've seen, most of the movement we need is collective, not individual. The question "should we incentivize the production of electric vehicles?" is more important than the question "should I drive an electric vehicle?" Whether you will, personally, purchase an EV depends in large part on its price, which in turn depends on collective, political decisions.

In this chapter, we'll first survey some fundamentals about policy. Then, we'll linger outside of the policy process by using another framework from the policy sciences to analyze the flow of how decisions are debated, made, implemented, and evaluated. Next, we'll immerse ourselves into the process as if we were the decision-makers to see what kinds of tools can help us decide. Finally, we'll look at carbon pricing and industrial policy as two major climate policy platforms.

What Is Policy?

Policy is a social process of authoritative decision-making through which members of a community (whose opinions and interests differ) clarify and secure their common interests (Clark 2002). Policy is the process by which a group decides: What are we going to do? Who shall get what, when, and how? Policymaking is inherently forward looking, because it is about selecting a path into the future.

We should note right away a special class of policy: Constitutive policymaking—like a nation's Constitution—establishes the framework for decisions. It is decisions about how decisions will be made. It answers the questions: Who gets to decide? By what process? It's helpful here to recall also the constitution of knowledge, which is a key way in which such questions get answered.

In summary: we are social creatures, so we must decide many things together. That's the basis of what we now call policy. Understood in these broad terms, policy is vital to the definition of a political community. The people in such a community share expectations about who has the authority to make decisions about what, when, and how. Indeed, faith in policy-making structures is really the glue that binds a civil polity together.

Policy is a term used in many ways: a plan, a program, a rule, a commitment to a course of action, case law, legislation, a contract, etc. It's important, then, to know what such words mean in different contexts, especially their legal standing and social force (e.g., do people actually follow the rules or are they ignored?).

It's also important to think about the distinction between policy and politics. There are various ways to parse this. For me, politics denotes the interplay of *special interests* that are inevitably involved in the policy process. There is no policymaking without politics, but they are not the same, because policy pertains to the process of finding *common interests*. Common interests are things that are good for the entire community like equal rights, clean drinking water, a safe food supply, and clean air. Our two main climate goals are common interests: climate safety and well-being. Special interests, by contrast, only benefit part of the community at the expense of others.

Of course, there isn't always a clear dividing line between special and common interests. Indeed, a great deal of policymaking entails wrangling about what a common interest *means* or what it *is* in a particular context and whether something is really a special interest masquerading as a common interest. This gets worked out through deliberation and compromise and will require context-sensitive judgments that call on our ethical reasoning and problem-framing skills. In any context, though, it will be helpful to ask of the policy process: is it open and inclusive, does it meet valid expectations, and is it flexible to meet goals as situations change? Evaluating the process with these criteria can help ensure that it stays focused on common interests.

In climate policy, we always face uncertainty (limits to our knowledge) and futility (limits to our control). As there is no "human society" outside of our multispecies entanglements, nature will often do a lot of the deciding for us. In the Colorado River case, for example, the drought is forcing the hand of decision-makers. If cuts are not made, Lakes Powell and Mead may get so low that water cannot flow downstream from the dams, creating a situation called "dead pool."

Human cultures have formed various ways to try to extend their knowledge and control. They might, for example, propitiate the gods and develop rituals to discern the inner workings of nature and to foretell the future. For the vast majority of human existence, these activities occurred on the small scales of family and tribe. The elders, shamans, or chiefs in various forms of counsel with their people would make decisions for the collective.

The rise of larger civilizations and empires necessitated the development of formal structures of power and governance. Most importantly for us, the coordinated decision-making capacities of European nations played a decisive role in their colonial conquests. As this global project of development has industrialized, it has become ever more bureaucratic and systematic in its decision-making structures. An ever-increasing range of human experience has become subject to planning, surveillance, and control. So much of our lives are shaped by policies.

The Decision Process: A Framework for Analyzing Policy

During the nineteenth and early twentieth centuries, there was an optimism that social scientists could achieve the kinds of precise control that natural scientists were starting to accomplish in their labs. Perhaps we could engineer society just as we engineer a machine, and maybe science could rationalize politics in the same way it rationalized chemistry and biology. This gave rise to visions of technocracy and hopes that science could cut through the tangles of interests to find "the one best solution" for any social problem.

We are humbler now. Recall our earlier discussions of wicked problems, hyper-objects, and paradoxes. We are not going to predict and perfectly control the multispecies drama of climate politics. And there are often incommensurable differences of values, such that there will never be agreement on what constitutes the "best" let alone a "good" solution. Yet we can still apply reason to our collective conduct and we can analyze these policy processes in ways that provide guidance so that we know where things stand, where we stand in relation to the whole, and where it may be most fruitful to intervene in the name of common interests. The policy sciences have already given us two tools for this: the problem orientation (Chapter 3) and the social process (Chapter 8). It's time to turn to the third part of this framework: the decision process (see Clark 2002).

The decision process is a tool for analyzing how a decision was made or is currently being made. It conceives of policymaking as consisting of seven functions or activities. These functions often interact, so they should not be

PolicyMaking as Consisting of Seven Functions or Activities

FIGURE 10.2

seen as strictly linear. Indeed, we have already critiqued the idea of a linear model of politics where science or "intelligence" leads straightforwardly to a common interest decision. You can even think of this model as nested within itself. For example, the invocation, application, and appraisal functions all entail intelligence gathering and promotion.

The key outcomes of any decision process are the policies, rules, or norms that a community is expected to follow. The consensus achieved does not mean everyone agrees with the resulting rules; it only means that (nearly) everyone expects these rules to be enforced by the community and its appointed officials.

Let's break down the seven functions into three groups: pre-decision, decision, and post-decision.

Pre-Decision

1. *Intelligence*: This is the process of obtaining and processing information about trends, conditioning factors, and projections relevant to the goals at stake in the policy. A great deal of the climate sciences studied in Chapters 5, 6, and 7 come to life here in the policy process.

In the Colorado River case, this entailed numerous state and federal agency reports on hydrology, water usage, the legal landscape, and more. Key questions you might ask about the intelligence phase of any climate policy, include: is the information credible, reliable, and relevant, and has it been made available to all stakeholders?

2. *Promotion*: This pertains to the activities of recommending and marshalling support for different policy alternatives. Another term for this would be advocacy. Here is where special interest groups make their case for what should happen with the Colorado River, usually arguing that they align with the common interests, say, of food, power, recreation, or wildlife preservation. What values are at play? How is the situation being characterized by different groups? How do different media outlets frame the issue? For example, is the Green New Deal described as a socialist plot or bonanza of green jobs? Recall here and throughout the decision process all of the participants and their values that we analyzed using the social process tool in Chapter 8 on climate politics.

Decision

3. *Prescription*: Sometimes called the anchoring decision, the prescription is most important. It establishes the rules by which people live, thereby crystallizing the demands and expectations of community members. In the Colorado River example, it is the decision to cut usage by three million acre-feet along with the specifics of who must forego which water uses. The content of a prescription must state its goals, the rules (how the goals will be achieved), the contingencies (situations in which the rules apply), sanctions (penalties for non-compliance), and assets (resources for applying the rules). Prescriptions must also have an authority signal—those who establish it must be recognized by the community as having the right and power to do so. And prescriptions must have control intent, that is, the authorities must be committed to continually upholding the rules. Without these elements, any decision will be more like a proposal—it won't be taken seriously and won't have prescriptive force. This should recall the "institutional inadequacy" that plagues international climate policymaking.

Post-Decision

4. *Invocation*: This is the first action taken by a community to put its prescription into effect. It is law (or rule) enforcement. When a community passes a speed limit ordinance, the police invoke it whenever they pull someone over for speeding. A game warden citing someone for illegal hunting or fishing is another example. Invocation entails administrative arrangements staffed by recognized authorities. For the Colorado River, it will entail local, state, and federal agencies ensuring that water users comply with the new rules. Questions to consider here include whether the implementation of the rules is done in a timely and fair manner and whether the sanctions are fair.

5. *Application*: This is the final decision on how a prescription will be implemented and who has the authority to do so. It entails interpreting the meaning of rules in a given situation with the aim of resolving disputes. The application function is usually carried out by courts, high-level regulatory commissions, or (in the case of private sector policy) chief executive officers. This function is usually prominent in climate policies, where many decisions end up in courtroom battles about the extent of federal and state/provincial control and the appropriate reach of government into the private sector.

6. *Appraisal*: The appraisal function assesses the policy process as a whole and the success of a prescription in achieving its goals. This will be vital in the case of the Colorado River as weather and climate trends will require constant evaluation of changing conditions. It's also crucial to consider the unintended consequences of the prescription. Indeed, unintended consequences are of utmost importance—recall that climate change itself is the unintended consequence of myriad policies. Policymakers are usually reluctant to undertake appraisals and are prone to skewing them favorably when it is their decision under scrutiny. So, it is important to ask who is doing the appraisal, by what standards, and with what information.

7. *Termination*: To terminate a prescription is to repeal, cancel, or significantly alter it. Ideally, this would occur after a fair appraisal process finds the prescription is failing to meet common interest

goals. What counts as termination may not always be clear cut. For example, the 2023 Colorado River prescription is partially a termination of the 1922 Compact. As with all aspects of the policy process, it is important to consider who will be harmed and who will be benefited when a prescription is terminated. Especially when a rule has been in place a long time, people may suddenly be deprived of benefits that they could previously count on or freed from regulations that had long burdened them.

In sum, looking at policymaking with these functions in mind can help you identify the important pieces of information, find the key players, and ask the right questions: What is the current or proposed practice or policy? Is it rational, practical, and morally justified? Where is this decision in the process? Are there important pieces of information or perspectives missing?

Picking Alternatives

The decision process offers a bird's eye view of policymaking to get our bearings. But now imagine swooping down into the fray. You are a decision-maker drowning in technical reports from the intelligence function and beset on all sides by advocates promoting their preferred alternative. Maybe you are deciding the allocation of river water, the approval of new transmission lines for a wind farm, the regulation of methane emissions from gas wells, or the fate of a subsidy for cattle ranchers. How do you decide?

You could just flip a coin or choose the path that best serves your personal gain. But set chance and corruption aside and assume that you care about more than just your own reputation, wealth, or power. How do you find the common interest as you listen in earnest to the stakeholders? The philosopher John Rawls (1971) offered a method of "reflective equilibrium" to organize the complex critical thinking that you might need to perform. He thought that a position could be justified by seeking coherence between theory, principle, and specific judgments.

This model of reflective equilibrium is pictured in Figure 10.3. The "normative ethical theories" listed at the top of this figure were discussed in the last chapter (consequentialism, deontology, and relational ethics). These are important for informing and correcting our intuitions about what makes something right or wrong, better or worse. But we also noted that they are

Reflective Equilibrium

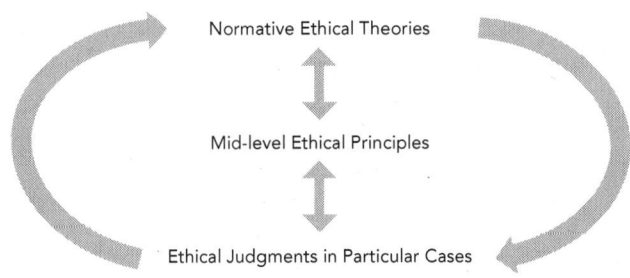

Normative Ethical Theories

Mid-level Ethical Principles

Ethical Judgments in Particular Cases

FIGURE 10.3

not sufficient for making decisions. Utilitarians and other consequentialists, for example, will disagree with one another depending on how they handle uncertainty, who they count in their calculations, and how they weigh competing goods. So, a top-down thinking needs to find some balance with a bottom-up thinking, which consists of drawing general insights from specific cases. This kind of case-based reasoning is central to law and policy.

Between the considered judgments on specific cases and abstract theory, there are mid-level principles. In biomedical and healthcare policy, for example, the principle of *informed consent* is often used in the middle level between theories of human rights and autonomy and specific cases of medical trials or practices. In environmental and climate policy, the *precautionary principle* is often invoked. Let's look at it briefly.

The Precautionary Principle

In the last chapter, it was shown how Hans Jonas argued that technological powers have changed the human condition, necessitating a new ethic. In his 1979 book, *The Imperative of Responsibility*, he formulated this maxim: "Act so that the effects of your action are compatible with the permanence of genuine human life." In so doing, he articulated a deontological theory of responsibility as sustainability (recall from the last chapter that deontology derives from the word "duty"). The *Vorsorgeprinzip* or precautionary principle developed out of this as a mid-level principle to guide specific actions in light of this more abstract imperative. The precautionary principle started to be used by German lawmakers in the 1970s to curb deforestation, air pollution, and other environmental harms.

The precautionary principle has since been given several formulations. Here are two of the most influential:

- 1992 Rio Declaration. "In order to protect the environment, the precautionary approach shall be widely applied by States according to their capabilities. Where there are threats of serious or irreversible damage, lack of full scientific certainty shall not be used as a reason for postponing cost-effective measures to prevent environmental degradation."

- 1998 Wingspread Statement. "When an activity raises threats of harm to human health or the environment, precautionary measures should be taken even if some cause and effect relationships are not fully established scientifically. In this context the proponent of an activity, rather than the public, should bear the burden of proof."

Note the common themes of (1) a responsibility for policymakers to anticipate and prevent harm before it occurs and (2) a reversal of the burden of proof for settling uncertainty such that the one proposing an activity must demonstrate it is safe, rather than the public demonstrating that it is dangerous. Some versions of that latter point suggest treating all technology as guilty until proven innocent.

Weaker versions of the precautionary approach are less controversial and quite common even in daily life. We regularly incur some costs to avoid future hazards even if they are unlikely to occur. Buying smoke detectors is a case in point. We tend to weigh the costs incurred against both the likelihood and severity of the possible harm. In this way, the precautionary approach is similar to cost-benefit analysis (below).

By contrast, strong versions of the precautionary principle hold that an activity should be prohibited if it has any potential significant adverse effects that are not fully understood, regardless of the costs imposed by such a prohibition. Even if an activity has tremendous possible advantages, if it has the slightest risk of irreparably damaging ecosystems, it should be banned. This version pits the precautionary principle against cost-benefit analysis, because it pictures our moral situation primarily in terms of a duty to avoid harming ecosystems, no matter what the cost. This is why the strong version is a mid-level principle that finds its equilibrium with the theory of deontology rather than consequentialism.

Despite its widespread application in climate and other forms of policy, the precautionary principle is not without its critics and problems. Max Moore (2004) for example, argues that the precautionary principle is itself a hazard to human progress. He formulates the "proactionary principle" as an antidote, claiming that the greatest risks we face are from nature, not technology, and that innovators need freedom to improve the human condition.

In another influential critique of the precautionary principle, Cass Sunstein (2005) argues that its weak version is trivially true (thus not useful) and that the strong version is incoherent. The key here is to note that precautionary policies themselves also carry risks of harm. So, if precaution is required in the face of uncertain harms, it negates itself by demanding and prohibiting action at the same time. There is no way to *not* decide—the choice for the status quo is still a choice, and it is one that carries its own risks. For example, banning genetically modified organisms can lead to increased hunger, prohibiting nuclear power can extend reliance on coal with its air pollution and climate risks, and suspending the use of a vaccine for fear of uncertain complications can cause preventable deaths from the virus.

So, the precautionary principle may be useful in calling attention to questions of uncertainty and burden of proof, but it hardly supplies a ready answer to vexed climate policy questions. It does seem to come down to a thoughtful weighting of risks, costs, and benefits.

Cost-Benefit Analysis

The most common method for picking alternatives is the cost-benefit analysis or CBA. This can be seen as a principle or decision procedure that is a mid-level step-down from consequentialist (especially utilitarian) theories. The simplest form of a CBA comes from Ben Franklin's "moral algebra," which consists of dividing a sheet of paper into two columns, listing all the "Pros" and "Cons" of an alternative, estimating their respective weights, and then comparing this with the same kind of accounting for the other alternatives. Just as Jonas's initial formulation of the precautionary principle came to be formalized, so too has Franklin's moral algebra. The field of economics has developed many refined CBA methods to find the point of economic efficiency where costs equal benefits at the margin.

An excellent example of CBA in climate policy is *Drawdown* (Hawken 2017), which is a solutions-oriented project to reducing GHG emissions and supporting natural sinks. Their library of 93 solutions is ranked in terms of benefits (tons of GHGs reduced or sequestered[1]) and costs (implementation and operations costs and savings). In other words, it lays out where we can get the most bang for our buck in terms of climate actions. For example, they estimate that increased onshore wind turbine deployment should be at the top of the list: it can cut up to 143 gigatons of CO_2 at a net savings of up to \$9.8 trillion.

At the heart of any CBA are the questions: What gets counted and how is it counted? Here there are four key concepts that we'll highlight with a little story. Imagine you fill up your car with a gallon of gas. Built into the price you paid on the pump are all sorts of market activities like the costs of extraction, transport, and refining of oil. But the price is also influenced by policy interventions known as *subsidies*: money given by governments to industries to keep the price of commodities low. The International Energy Agency estimates that fossil fuel subsides topped \$1 trillion in 2022 (IEA 2023).

As you drive around, you burn that gallon of gas, emitting GHGs. Those emissions contribute to climate change and cause damage, but you didn't pay for those damages. They were not reflected in the price of the gasoline. That's the next concept, a *negative externality*: The effect of a commercial or industrial activity that negatively affects other parties but is not included in the costs of the goods or services involved. (An example of a *positive externality* from a different area is the pollination of nearby crops by bees kept for honey.) In a way, this negative externality is a subsidy, after all, it keeps the price of gas lower.

Lots of climate policy analysts have tried to quantify those externalized costs of carbon dioxide and other GHGs. The ethical idea is simple: if you impose a cost on society through your consumption, you should pay for it. The cost should be internalized in the price of the commodity. That cost is just a quantification of the externalized harm and it goes by the name the *social cost of carbon*. If we could figure that out, we could, say, impose a tax on gasoline and other sources of GHG emissions. This would give us a more accurate CBA and steer behavior toward our climate goals.

The climate sciences can give us rough estimates of how much damage a ton of carbon does. It is a trickier matter of economics and ethics to determine how much it is worth for us to avoid that damage. We saw with "the perfect moral storm" of climate change that there is a dispersion of cause

---

1   Confined and isolated. Carbon sequestration removes carbon from the atmosphere and stores it.

and effects—many of the damages from today's GHG emissions will come years later. The policy challenge, then, "is to pull those damages out of the future and into the present" (Roberts 2012). This gets us to the *discount rate*: the amount (expressed as a percent) that a benefit declines in value each year that it is pushed back into the future.

The discount rate that we choose tells us how much the social cost of carbon is. The UK economist Nicholas Stern (2007) used a 1.4 percent discount rate to estimate the social cost of carbon at $311/ton and rising. This means that climate policies today are justified even if they cost $311 per ton of GHG emissions prevented. By contrast, using much the same scientific data, the US economist William Nordhaus (2007) replied that 3 percent is a more reasonable discount rate. Using that rate, Nordhaus calculated the social cost of carbon at just under $20/ton.

Once again, the moral of the story is that facts matter but facts don't speak for themselves. The discount rate is not the only place where judgment and interpretation come into play. Estimates for the social cost of carbon also rely on assumptions of climate sensitivity, emissions scenarios, and other aspects of climate science. One crucial element here pertains to the assumptions made about the costs of renewable energy. Traditional models (like the IAMs used in much climate forecasting) have long underestimated the rapid price declines of solar and wind power. These technologies are on very steep "learning curves" so the price is dropping rapidly as implementation progresses. This may continue, meaning that a rapid energy transition won't be an overall cost to the global economy; rather, it will be a net gain (see Way et al. 2022).

Climate Policy Platforms: Carbon Pricing and Industrial Policy

We can also examine climate policy from a macro-level as the platform that a nation or organization takes as the basic structure to their activities. Platforms set the overarching agenda and tone for policymaking. Let's recall our two main climate policy goals: 1) safety or avoiding "dangerous anthropogenic interference in the climate system" and 2) promoting human well-being. These have largely been characterized in terms of 1) net zero emissions by 2050 or 1.5°C of warming and 2) continued economic growth usually measured by GDP. Consider also Figure 10.4, which shows that despite the rapid growth in renewables, we've actually made very little progress in decarbonizing the energy system in the decades since climate policies started being formulated.

Global Energy Consumption by Source

Primary energy consumption is measured in terawatt-hours (TWh). Here an inefficiency factor (the 'substitution' method) has been applied for fossil fuels, meaning the shares by each energy source give a better approximation of final energy consumption.

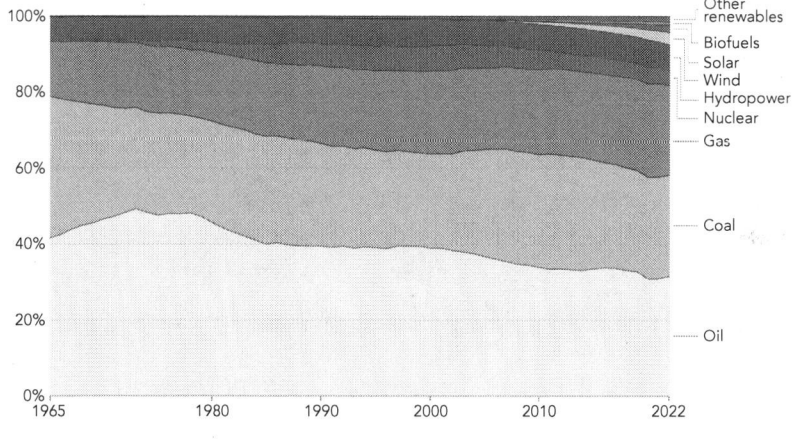

Note: 'Other renewables' includes geothermal, biomass, and waste energy.

FIGURE 10.4

There is no silver bullet policy solution to climate change. Different actors (nations, cities, corporations, etc.) are trying out various mixes of policy alternatives to achieve these goals in a dynamic balance with other goals, especially profit and economic growth. We can roughly organize much of this activity, though, into one or another of two major policy platforms: carbon pricing and industrial policy. These approaches are sometimes at odds with one another, but many nations deploy aspects of both. We might think of carbon pricing as the stick (making fossil fuels more expensive) and industrial policy as the carrot (making alternatives cheaper).

Carbon Pricing

As many economists argue, the most efficient way to achieve climate mitigation goals is to tax carbon. Thus, lots of CBA work in climate policy focuses on carbon pricing. It makes sense ethically: people should pay for harms they cause. And it makes sense pragmatically: money is a principal driver of behavior, so if the price of fossil fuels goes up, people will shift to alternatives. Several European nations, Canada, China, and others have carbon pricing policies.

There are two broad categories of carbon pricing, both aimed at internalizing the externalized or social cost of carbon:

- *Carbon tax*: These are taxes levied on the carbon (and sometimes other GHG) emissions required for producing goods and services. To keep such taxes from being regressive (imposing a greater burden on the poor rather than the rich), nations can redistribute the revenues generated by the tax to low-income groups. Such policies are often referred to as a "carbon fee and dividend." Carbon tax policies also often address concerns about "carbon leakage" (where emissions shift to areas with lower regulations) and possible loss of competitiveness (e.g., by imposing tariffs on goods imported from nations that do not have a carbon tax).

- *Cap-and-Trade* (CAT) or *Emissions Trading Schemes* (ETS): these are policies that create a market with limited allowances for emissions. Rather than setting a price per ton of carbon (as with a tax), the cost of carbon automatically adjusts to the cap that has been set. Polluters with emissions higher than their quota can purchase the right to emit more from others who choose to stay under their quota. A key weakness of ETS is their vulnerability to forms of cheating like unverified or non-existent emissions reductions (an issue we'll return to in the next chapter). A key strength is their incorporation of market dynamics, which often makes them more politically palatable.

Carbon pricing has not worked well in the US, the world's largest cumulative emitter of GHGs. In the 1990s, the Clinton Administration proposed a BTU tax, that is, an energy tax on fossil fuels. The fossil fuel lobby successfully killed the tax. Carbon pricing was resurrected in the first term of the Obama Administration with the Waxman-Markey CAT bill. This attempt also failed, again leading to years of federal drift on climate policy.

Industrial Policy

Nonetheless, by 2020 the US actually surpassed the emissions targets that had be set by the failed CAT proposal. Efforts in this direction were possibly motivated by the protests of the youth-led Fridays for the Future and Sunrise Movement. Then came the Green New Deal, a holistic vision of a new economy centered on the intersectionality of social and climate justice. Environmental NGOS found new life and big ideas, including the Blue-Green Alliance, which brings together environmentalists and labor.

Along with this general momentum came what the policy analyst David Roberts (2020) called a paradigm shift in the US climate policy platform. Carbon pricing is no longer at the center of policy ideas. Rather than focus on punishing fossil fuels, the emphasis is on making alternatives so cheap that people will willingly flock to them. The various climate rebates, subsidies, and incentives in the massive 2022 US Inflation Reduction Act are an excellent example of climate industrial policy.

The key to industrial policy is using the government to guide economic growth in a particular direction by passing laws and regulations that boost renewable energy and other climate-friendly industrial activities. This is not *laissez faire* capitalism, but it is still capitalism, because the main engine is private industrial activity and the direction is economic growth. It's just guided by strong government interventions with the rationale that low carbon energy for everyone is a common interest (see Nordhaus 2019). In the US, industrial policy hearkens back to the vision of Alexander Hamilton and has been a part of several massive economic booms: the transcontinental railroad, the suburb, home appliances, and the computer (see Meyer 2019). The push now is for wind farms, solar panels, electric vehicles, heat pumps, better batteries, and other technologies needed in the race toward net zero.

Roberts argues that there are three main pillars to this approach, which he calls SIJ:

* *Standards*: Sector-specific performance standards designed to drive GHG emissions out of power generation, transportation, buildings, and industry as quickly as possible. Examples include requirements for electric vehicle manufacturing and renewable electricity on the grid (the latter is often called Renewable Portfolio Standards).

- *Investment*: Lage scale public investments in green industries, technological innovation, research, and green jobs. These are subsidies that speed up adoption curves and drive down costs of renewables.

- *Justice*: This means prioritizing the needs of labor unions, fossil fuel workers, and vulnerable front-line communities.

Rather than a single fix, the climate journalist Robinson Meyer (2021) argues that industrial policy creates a virtuous cycle or "the green vortex." Policy speeds up development, which makes green technologies cheaper, which means more companies adopt them, which means they are more amenable to climate policies that favor their bottom line, which means coalitions shift to bring government, labor, and corporations into alignment, which fuels the cycle further.

Of course, the path ahead is still daunting as the US and most of the world remain off target to reach climate goals. A global energy transition is a hard thing to rush. It may be that in addition to carbon pricing and industrial policy we will need more supply-side policies to keep carbon in the ground by explicitly banning extraction and the construction of new fossil fuel infrastructure. This could, for example, take the form of changes to permitting policies that effectively outlaw fossil fuel extraction. The emerging Fossil Fuel Non-Proliferation Treaty is an effort by climate organizations to pressure governments to commit to ending new oil, coal, and gas projects.

Conclusion

In this chapter, we picked up some tools both for analyzing policy formation and for guiding decisions in the midst of that process. It is remarkable that such diverse civil societies all participating in the grand "development project" are able to hold themselves together around the policy processes through which they clarify and secure their common interests. Climate change poses enormous challenges to these processes because of the speed and scale of changes required to systems that are backed by powerful special interests. In the next chapter, we'll zoom in from these general policy considerations to examine the three main types of climate policy alternatives: mitigation, adaptation, and finance.

Activities and Questions

1. Research the Fossil Fuel Non-Proliferation Treaty. What is their mission? Do you think this is a promising approach to climate policy— why or why not?

2. Why do you think carbon pricing works in some nations but not others? What social, cultural, or political factors influence the acceptability of different climate policies?

3. What connections can you make between green growth and industrial policy? What might a proponent of degrowth say about industrial policy?

4. Do you think the precautionary principle makes sense or is helpful for guiding decisions—why or why not? See if you can find a case where people have drawn vastly different conclusions from CBA approaches— what does this say about the interpretive flexibility of numbers and reality?

References

Clark, Susan. 2002. *The Policy Process: A Practical Guide for Natural Resource Professionals*. New Haven, CT: Yale University Press.

Hawken, Paul, ed. 2017. *Drawdown: The Most Comprehensive Plan Ever Proposed to Reduce Global Warming*. New York: Penguin.

International Energy Agency. 2023. "Fossil Fuels Consumption Subsidies 2022." February. https://www.iea.org/reports/fossil-fuels-consumption-subsidies-2022.

Jonas, Hans. 1979. *The Imperative of Responsibility*. Chicago: University of Chicago Press.

Jones, Benji. 2023. "Why the Colorado River Agreement Is a Big Deal— Even If You Don't Live Out West." Vox, May 23. https://www.vox.com/climate/2023/5/23/23734404/colorado-river-cuts-lake-mead-deal.

Meyer, Robinson. 2019. "A Centuries-Old Idea Could Revolutionize Climate Policy." *The Atlantic*, February 19. https://www.theatlantic.com/science/archive/2019/02/green-new-deal-economic-principles/582943/.

—. 2021. "How the U.S. Made Progress on Climate Change without Ever Passing a Bill." *The Atlantic*, June 16. https://www.theatlantic.com/science/archive/2021/06/climate-change-green-vortex-america/619228/.

Moore, Max. 2004. "The Proactionary Principle." Extropy Institute. https://www.extropy.org/proactionaryprinciple.htm.

Nordhaus, Ted. 2019. "The Empty Radicalism of the Climate Apocalypse." *Issues in Science and Technology* 35, no. 4 (Summer): 69–78. https://issues.org/empty-radicalism-of-the-climate-apocalypse/.

Nordhaus, William. 2007. "A Review of the Stern Review on the Economics of Climate Change." *Journal of Economic Literature* 45 (3): 686–702.

Rawls, John. 1971. *A Theory of Justice.* Cambridge, MA: Belknap Press.

Roberts, David. 2020. "At Last, a Climate Policy Platform That Can Unite the Left." Vox, July 9. https://www.vox.com/energy-and-environment/21252892/climate-change-democrats-joe-biden-renewable-energy-unions-environmental-justice.

—. 2012. "Discount Rates: A Boring Thing You Should Know About (with Otters!)." Grist, September 24. https://grist.org/article/discount-rates-a-boring-thing-you-should-know-about-with-otters/.

Shao, Elena. 2023. "See How the Colorado River Water Gets Used Up." *New York Times*, May 22. https://www.nytimes.com/interactive/2023/05/22/climate/colorado-river-water.html.

Stern, Nicholas. 2007. *The Economics of Climate Change: The Stern Review.* Cambridge: Cambridge University Press.

Stone, Deborah. 1998. *Policy Paradox: The Art of Political Decision Making.* New York: W.W. Norton.

Sunstein, Cass. 2005. *Laws of Fear: Beyond the Precautionary Principle.* Cambridge: Cambridge University Press.

Way, Rupert, et al. 2022. "Empirically Grounded Technology Forecasts and the Energy Transition." *Joule* 6 (9): 2057–82. https://doi.org/10.1016/j.joule.2022.08.009.

CHAPTER 11

Mitigation, Adaptation, and Finance

Simplified Emissions Pathways for Climate Targets

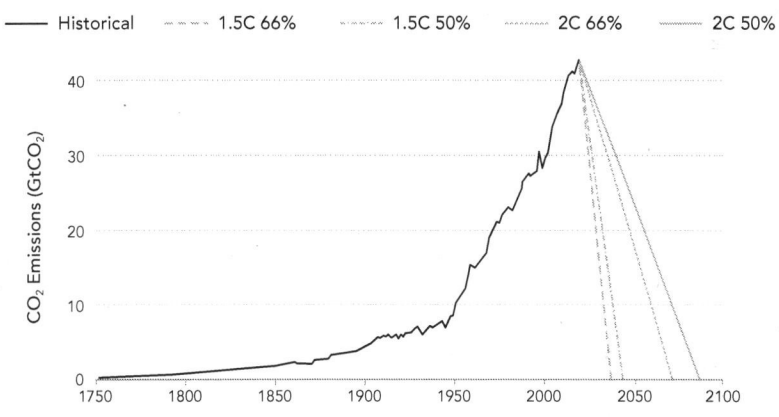

FIGURE 11.1

Figure 11.1 pertains to mitigation and illustrates how daunting the task is to reach our 1.5°C or 2°C targets. But it is even more challenging than this graph depicts, because the global economy is expected to keep growing. Long-term economic growth is subject to deep uncertainty, as is reflected in the divergent climate model results from different scenarios (see Burgess et al. 2023). It seems like a good estimate is about a four-fold increase in per capita GDP between 2000 and 2100. It's one thing to decarbonize the present economy, but how will we decarbonize an economy that is many times larger?! That's a key question for this chapter.

Meanwhile, all those emissions are driving increased frequency and intensity of some weather and climate extremes. Most of the world's infrastructure was built for the Holocene climatic conditions that we are rapidly leaving behind. So, protecting ourselves in this emerging new climate will require a massive global construction project (Wallace-Wells 2022). How can this be done? That's a matter of adaptation, and it is the other major question for this chapter.

A big reason storm damage costs are rising is because people are moving to and building infrastructure in the expanding bullseyes of extreme weather. In the US, like many parts of the world, the typical response to weather disasters is to rebuild in the same place in much the same way. That, however, is starting to change. The federal government is creating programs for "managed retreat," helping communities in some of the most vulnerable areas relocate elsewhere.

One of the first programs was created in 2021 when Congress provided the Bureau of Indian Affairs (BIA) with $130 million to help Native American tribes to relocate. Tribes can apply for grants up to $3 million/year. For tribes, this is a new chapter in a long history of involuntary migration—first the forced relocation by conquest and now managed retreat. Indeed, the historic acts of dispossession that confined tribes to marginal lands are a huge factor behind their extreme vulnerability to climate change today.

The Shoalwater Bay tribe in Washington state, for example, is rapidly losing parts of its reservation in a "double whammy" of increased inland flooding and coastal surges (Flavelle and Irvine 2022). Since 2013, the US Army Corps of Engineers has spent millions constructing artificial dunes to try to keep the broad coastal marshlands of the reservation from eroding. But between storms and rising seas, the dunes keep failing. The tribe needs to move to higher land.

And here we get to our other major question, which is about finance: where does the money come from? The $3 million grant isn't even enough to build a road to the new location scouted out by the Shoalwater Bay tribe, let alone actually build homes and other infrastructure. For all of that, they need closer to $500 million. With such scarce resources, the toughest question faced by the BIA is who to help first. There are 574 tribes in the US, many of which will need assistance in the face of climate impacts. The funds available are orders of magnitude too small. And this doesn't even account for the needs of non-native people, like the residents of Santa Cruz, California whose roads and homes are also crumbling into the Pacific.

Zooming out from this microcosm of the challenges for climate policy, let's recall how we framed our goal of climate safety in Chapter 3: Reduce the risks and impacts of anthropogenic climate change through mitigation (preventing and minimizing dangers by reducing GHG emissions) and adaptation (increasing security and resilience to withstand climate dangers). This chapter explores these areas in more depth through a survey of some of the key concepts, issues, and debates surrounding the intertwined tasks of mitigation, adaptation, and finance.

Mitigation

We know in the UNFCCC process that each nation attempts to mitigate their own emissions. Yet, corporations, states, provinces, cities, and even college campuses are also increasingly adopting climate plans to tackle the problem at different scales. Indeed, cities are some of the most active hubs for climate mitigation and adaptation work (see Miller 2020). In this section, we'll focus on four key aspects of mitigation: a) the fundamental options; b) emissions by sector and the goal to electrify everything; c) skepticism about the energy transition; and d) the allure of carbon offsets.

The Fundamental Options

If the goal is to reduce emissions, it is helpful to ask where they come from or why the emissions exist in the first place. What are the drivers of GHG emissions? It's too specific to say something like "tailpipes" or "smokestacks" and it's too abstract to say something like "humans" or "a consumer capitalist vision of the good life." One popular model is IPAT, or for our purposes EPAT (where the E stands for emissions rather than an I for impact):

$$\text{Emissions} = \text{Population} \times \text{Affluence} \times \text{Technology}$$

A slightly more refined expression is the Kaya Identity.

The Kaya Identity

$$\text{GHG} = \text{Pop.} \times \frac{\text{GDP}}{\text{Pop.}} \times \frac{\text{Energy}}{\text{GDP}} \times \frac{\text{GHG}}{\text{Energy}}$$

FIGURE 11.2

This shows us that GHG emissions are driven by population, per-capita wealth (GDP/Pop.), energy intensity (Energy/GDP, the amount of energy it takes to create a unit of wealth), and the GHG (or carbon) intensity (GHG/ Energy, the amount of emissions it takes to create a unit of energy). It is called an "identity," because the terms on the left and right side of the equation are the same (you can cross out all the numerators and denominators and be left with GHG = GHG).

The Kaya Identity tells us that there are four main levers we can pull to mitigate emissions. We can reduce:

1. human population;
2. the amount of wealth each person has;
3. the energy intensity of the economy to create more wealth with less energy; and/or
4. the GHG intensity of the energy that we use (including food calorie energy).

Figure 11.3 presents this idea in graphic form with empirical data. You can see that, globally, energy intensity and carbon intensity are declining, but not as rapidly as population and per capita wealth are increasing, which means that overall emissions have continued to rise.

For lever one, it is important to note the dehumanizing and violent track record of some approaches to population control. However, many argue that bending the population curve can and should be done in ways consistent with human dignity and well-being, by focusing on education, women's empowerment, access to contraceptives, and new forms of kinship (see Tucker 2019). We surveyed the debates around degrowth, lever two, in Chapter 4. Of course, most of the attention in the predominant climate discourse focuses on lever four, the carbon or GHG intensity of energy via the transition to wind, solar, and perhaps nuclear, hydrogen, and geo-thermal. Lever three (energy intensity, the energy required to produce one unit of wealth) is basically a measure of economic efficiency, which raises the potential for *Jevon's Paradox*, which states that an increase in resource efficiency creates an increase, rather than a decrease, in consumption. The cheaper the energy, the more we tend to use. So, it is unclear to what extent energy efficiency gains will equate to long-term GHG emissions reductions.

The Kaya Identity and Global Drivers of CO_2 Emissions

Percentage change in the four parameters of the Kaya Identity, which determine total CO_2 emissions. Emissions include fossil fuel and industry emissions.[1] Land use change is not included.

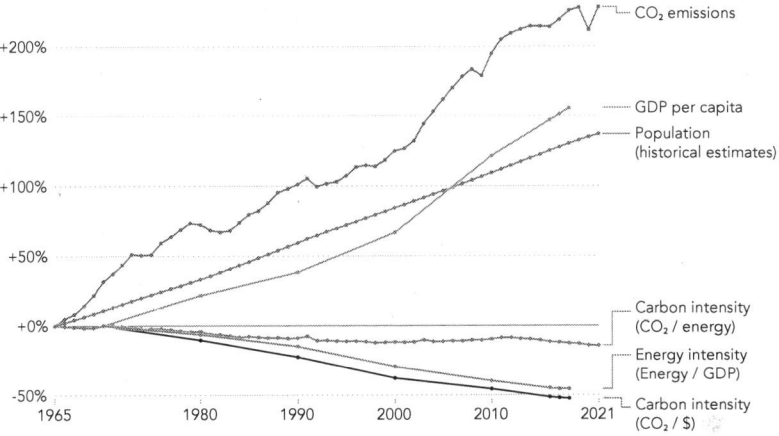

Note: GDP per capita is measured in 2011 international-$[2] (PPP). This adjusts for inflation and cross-country price differences.

1. Fossil emissions: Fossil emissions measure the quantity of carbon dioxide (CO_2) emitted from the burning of fossil fuels, and directly from industrial processes such as cement and steel production. Fossil CO_2 includes emissions from coal, oil, gas, flaring, cement, steel, and other industrial processes. Fossil emissions do not include land use change, deforestation, soils, or vegetation.

2. International dollars: International dollars are a hypothetical currency that is used to make meaningful comparisons of monetary indicators of living standards. Figures expressed in international dollars are adjusted for inflation within countries over time, and for differences in the cost of living between countries. The goal of such adjustments is to provide a unit whose purchasing power is held fixed over time and across countries, such that one international dollar can buy the same quantity and quality of goods and services no matter where or when it is spent. Read more in our article: What are Purchasing Power Parity adjustments and why do we need them?

FIGURE 11.3

Emissions by Sector and Electrify Everything

Another way to analyze emissions is by sectors of the economy. It is common to break emissions down into sectors or categories like transportation, buildings, agriculture, and power generation. There are many ways to divide the emissions pie. For example, one study estimated that the world's militaries taken together as a sector are responsible for 5.5 percent of global GHG emissions (Parkinson and Cottrell 2022). If the world's militaries were a country, it would have the fourth highest carbon footprint. Figure 11.4 offers one way to breakdown the sources of emissions in the global economy.

Clearly, there is a wide variety of sources, meaning that there is no single solution for mitigation. Yet there is something like a master plan: electrify everything (see Griffith 2022). This is increasingly the consensus among climate and energy experts. The idea is to continue pushing for a

Global Greenhouse Gas Emissions by Sector

This is shown for the year 2016 – global greenhouse gas emissions were 49.4 billion tonnes CO_2 eq.

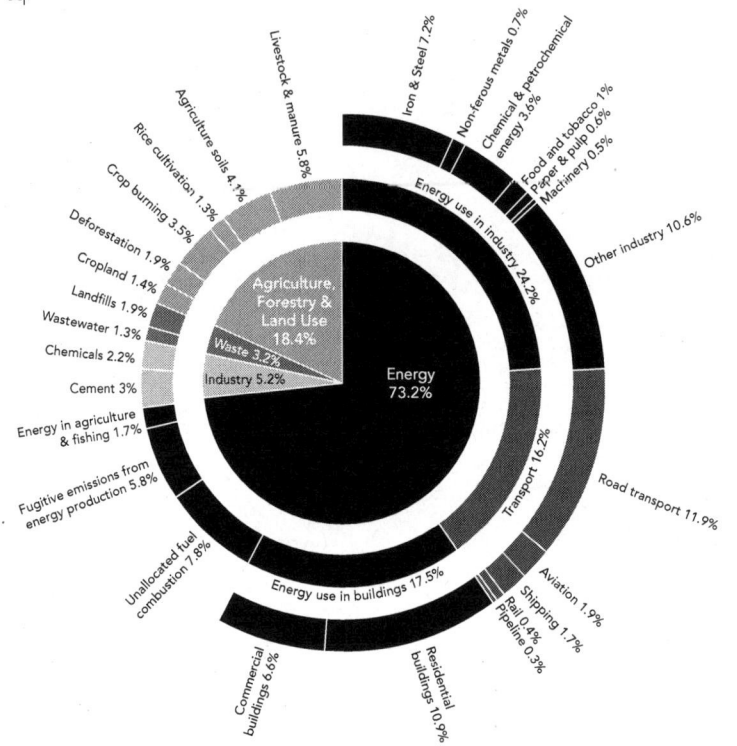

FIGURE 11.4

100 percent carbon free grid and replace all technologies that run on combustion (gasoline vehicles, natural gas building heating, industrial smelters, etc.) with alternatives that run on that renewably generated electricity (EVs, heat pumps, etc.).

Figure 11.5 can illustrate this. It is a Sankey flowchart or diagram of US energy consumption. The idea is to replace the natural gas, coal, and petroleum sources with more renewables and nuclear power. In addition, the transportation, industrial, commercial, and residential uses would be shifted over to electricity from the cleaner grid. This figure also demonstrates just how inefficient the energy system is: 67.3 percent of all energy produced is "rejected," meaning it is wasted. Only 33 percent goes toward useful "energy services" like moving vehicles, powering computers, and cooling homes.

Estimated US Energy Consumption in 2022

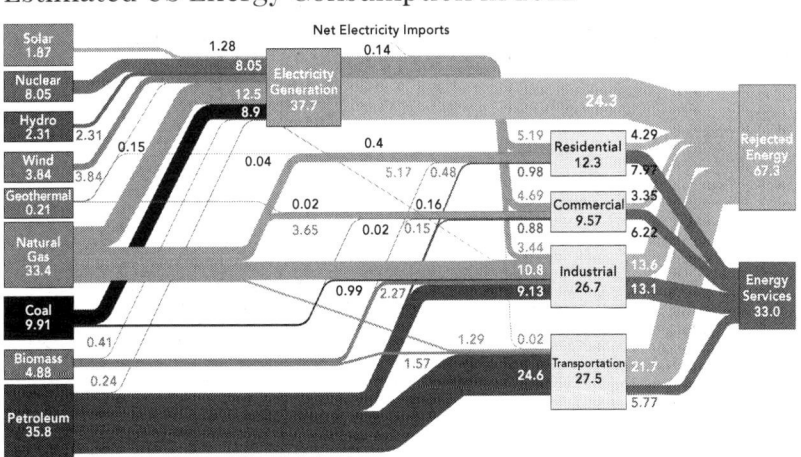

FIGURE 11.5

Built into the goal of electrifying everything are not just the technical challenges, but also the social and political ramifications that come with the shifting material foundations of the energy transition. Some of the political action is local, for example, with decisions about siting and constructing new infrastructure. Other aspects of the politics of "electrifying everything" are global. A movement from coal and oil to lithium and cobalt will bring new international and corporate alignments, perhaps ushering in an age when new "electro-states" displace the old "petrostates." Shifts are already happening as nations seek to secure supply chains through various means, including onshoring of extraction, processing, and manufacturing when possible.

Of course, "electrify everything" doesn't capture all the mitigation work that needs to be done. Another major problem stems from methane emissions, which come primarily from waste (i.e., landfills), leaks in the oil and gas industry (called fugitive emissions), as well as animals raised for food. Methane is roughly thirty times more powerful than CO_2 when it comes to trapping heat. This is why treaties like the Global Methane Pledge (launched at COP 26 and given more than $1 billion in additional funding at COP 28) are so vital. If the 150 nations signed onto this pledge meet their commitments, they could avert 0.2°C of warming.

Skepticism about the Energy Transition

Mitigation is clearly an important goal to pursue for all of the reasons we have explored. It is not, however, without its tradeoffs and risks. No machines, even those powered by renewable electricity, are without environmental and social impacts. So, it is important to be aware of the costs and tradeoffs imposed by the great energy transition.

There are many factors that warrant questioning, and there are different degrees of skepticism. On the lighter side of the spectrum, of course, is the common argument (well-supported by the IPCC) that the transition is not happening fast enough to reach our goals. This is for a variety of reasons from economics (finance!) to physics (the lower power density[1] of renewables) to social and political factors (for example, denial and delay tactics by the fossil fuel industry). Taken together, this means that even countries that invest heavily in renewable energies will still likely need fossil fuels for a long time. China, for example, is the world's largest and fastest-growing producer of renewable energy. However, despite years of growth in renewable power installations, in 2023 China still generated roughly 70 percent of its electricity from fossil fuels.

These lighter skeptics might also note that all the growth in renewables is so far largely additive—just piling on top of a stubborn fossil fuel base, rather than replacing it (the example of China also applies here). And they will point to the environmental and human rights costs of renewable energy technologies, including the brutal mining practices for cobalt in the Democratic Republic of Congo (Kara 2023). They will debate what policies are needed to rectify the problems (more rare metal recycling, stricter labor laws, etc.), but they tend to agree that the basic plan of green growth is sound.

Toward the heavier side of the skepticism spectrum, though, positions get more radical as we shift into degrowth territory. Maybe the idea of replacing a complex industrial system powered by fossil fuels with an even more complex system powered by the intensive mining of several rare earth materials is a fool's errand. The journalist Andrew Nikiforuk (2023) surveys some of the key skeptics making this point. They argue that there are not enough minerals to power a future civilization on renewable energy sources. Even if there were, the social, banking, and political infrastructures needed to mobilize and coordinate all the required activities are cratering

---

1   The amount of power per unit volume.

under declining trust, geopolitical conflict, and the beginnings of climate disruptions. In short, there are hard limits to growth and it is time to consider how to build a simplified future with a smaller footprint and alternative social structures. This is what we noted earlier is sometimes called "deep adaptation." It means building a very different future world rather than a world that looks much like ours but is powered entirely by renewable energy.

Two notes of caution are in order. First, there are good faith and reasonable skeptical positions, but be careful to distinguish these from fossil fueled propaganda. Illegitimate climate denial and delay often drives bad faith skepticism about mitigation. Second, the environmental costs of the energy transition to a "green economy" need to be compared with those imposed by our current fossil fueled world. For example, electrifying everything will require lots more lithium and cobalt for batteries, but this needs to be put into the wider context. In 2021, about 170,000 tons of cobalt were extracted, which is a tiny fraction of the 2.6 billion tons of iron ore. The 7.9 billion tons of coal extracted is about 74,000 times as much as the lithium extracted. Of course, such comparisons are complicated due to resource abundance, metal-to-ore ratios, and other technical issues. But the point is that our fossil-fueled economy is already built on extreme extractivism.

The Siren Song of Carbon Offsets

Carbon offsets are the reduction or removal of emissions of CO_2 or other GHGs made in order to compensate for emissions made elsewhere. A carbon credit is the certified financial instrument to represent this reduction or removal. The basic idea is that you pay someone to remove the equivalent amount of GHGs that you added to the atmosphere, perhaps by planting trees or by funding renewable energy. This is sometimes done as part of a mandatory cap-and-trade system (Chapter 10), but there is also a voluntary offset market. These offsets are even available for individual consumers (for example, paying a little extra to offset the emissions from your flight or your package delivery).

By 2021, one-fifth of the world's 2,000 largest publicly listed companies had committed to a net zero emissions target, including United Airlines, Shell, and Apple (Graham 2021). Most of these plans rely on offsets rather than emissions reductions, even though numerous studies show that most offsets do not reliably reduce emissions (White and Rathi 2022). This is because offsets are easy to promise and sound great for marketing purposes: "We're carbon neutral!" But it is often hard to prove offsets are actually happening, and there are not many regulations in place to prevent abuse, for example, of the requirement that the carbon removed by the offset would otherwise not have been removed. So, the offset would have to plant a tree that would not otherwise have been planted or save a tree that would otherwise have been cut down. But companies have claimed they are carbon neutral by "preserving" forested land that is not actually under any threat or by claiming credit for financing a wind farm that would have been built anyway. And of course, even if a forest was legitimately protected, it may burn down, releasing its carbon into the atmosphere.

So, some healthy skepticism is warranted around offsets, which frankly often do more harm than good. Buyers can easily make big claims of being carbon neutral and sellers can make money for doing very little (like just not cutting down trees they weren't going to cut down anyway). The result is mere greenwashing—the appearance of climate action without any substance. The key is to scrutinize the carbon offset registries that serve as neutral third parties to verify the authenticity of the offset. In policy-speak, this goes by the phrase MRV: Monitoring, Reporting, and Verifying. Oftentimes, MRV is inadequate and the parties involved are not held accountable by any government.

Adaptation

Though adaptation and mitigation are sometimes pitted against one another as competing for scarce resources, they are often complementary. In my city of Denton, Texas, for example, we are making a Climate Action Plan and many of the mitigation measures to reduce emissions, like improved building envelopes for homes and businesses, also help us adapt to a warming climate.

The IPCC defines adaptation as "the process of adjustment to actual or expected climate and its effects." Note that this means adaptation measures can be sensible policy regardless of attribution—even if a weather hazard is not attributed to greenhouse gas emissions, it is still a hazard and it still makes sense to reduce vulnerabilities (see Pielke 2022). Arguably, in the context of loss and damage (Chapter 9) and other areas of climate policy, we get too hung up on attributing the source of hazards. Do we really need single-event attribution studies to justify aid for adaptation? Don't we know enough already to see that wealthy nations have an obligation to help poorer nations?

Following the structure of the IPCC's 2022 report on impacts, adaptation, and vulnerability, here we'll focus on four key areas: a) the definition of risk and other central terms; b) observed and projected impacts and risks; c) adaptation strategies; and d) climate resilient development.

Defining Risk and Other Central Terms

Adaptation is a form of risk assessment and management. Because safety is understood as reducing risks and impacts, it's important to get clear on these key terms. In a special guidance report on the concept of risk, the IPCC (2020) defined it as, "The potential for adverse consequences for human or ecological systems, recognising the diversity of values and objectives associated with such systems" (4). The "adverse consequences include those on lives, livelihoods, health and wellbeing, economic, social and cultural assets and investments, infrastructure, services (including ecosystem services), ecosystems and species" (4). Risks can arise both from potential impacts of climate change as well as human responses to it.

The hazards, exposure, and vulnerability are all subject to uncertainty in terms of magnitude and likelihood of occurrence (the uncertainty or the "potential" for bad things to happen is why it is called "risk"). The hazard is the adverse event that might occur. Exposure is the presence of people, livelihoods, ecosystems, species, infrastructure, and other valuables that could be adversely affected. And vulnerability is the propensity to be adversely affected, which relates to properties like susceptibility to harm and the capacity to withstand, cope, and adapt.

For example, a hurricane (hazard) approaches a city (exposure) with its building standards, warning systems, etc. (which determine its vulnerability). It's helpful to think of risk as the overlapping area in a Venn diagram as illustrated in Figure 11.6.

Finally, there are also psychological dimensions of adaptation, which we often discuss as "getting used to" a "new normal." A key concept here is *shifting baseline syndrome*: "In the absence of past information or experience with historical conditions, members of each new generation accept the situation in which they were raised as being normal" (Soga and Gaston 2018, 222). What Rachel Carson called a "silent spring" in 1962 due to the absence of song birds from pesticides, might strike later generations as simply a

Risk Venn Diagram

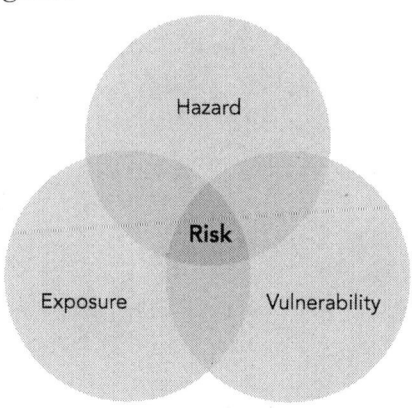

FIGURE 11.6

normal spring. Consider the collapse of biodiversity (Chapter 1), which is initially experienced as an impoverishment but over time that sense of loss fades as each generation redefines the "natural" or "normal" baseline. It is a kind of generational or cultural amnesia. The once unthinkable can become a daily reality to which one adapts. In the *House of the Dead*, Fyodor Dostoevsky offers a definition of humans as the "creature that can get accustomed to anything." It's fascinating to think of this as a paradox—as both our strength and weakness, as both a triumph and a tragedy.

Observed and Projected Impacts and Risks

The IPCC uses "impacts" and "consequences" interchangeably, noting that they can be adverse or beneficial. (So, risk is defined as the potential for adverse impacts.) They describe and quantify adverse impacts in terms of damages, harms, and economic and non-economic losses. They use the tools we discussed in our section on climate sciences (Chapters 5–7) to characterize observed risks and to project risks based on different scenarios. The IPCC also highlights complex risks where multiple hazards combine and influence interacting social and ecological systems.

The picture painted by the IPCC (2022) is complex, because risks and impacts differ for each hazard. The IPCC does, however, reach many conclusions about impacts with "very high" or "high confidence," including:

* Climate change has adversely affected physical and mental health globally.
* The rise in climate and weather extremes has led to some irreversible impacts, pushing some human and natural systems beyond their ability to adapt.
* Observed impacts include warm-water coral bleaching and mortality, drought-related tree mortality, increased areas burned by wildfires (medium to high confidence), substantial and irreversible losses to terrestrial, freshwater, and ocean ecosystems, loss of hundreds of local species, increases in mass mortality events, increased permafrost thawing, and decreased food and water security.
* Infrastructure (transportation, water, sanitation, and energy) has been compromised and resulting impacts are concentrated among economically and socially marginalized people.
* Climate and weather extremes are increasingly driving displacement in all regions.

Similarly, exposure and vulnerability are complex, because they greatly differ among and within regions. Here too, though, the IPCC does make several conclusions with "high confidence," including:

* Levels of vulnerability are driven by historic and ongoing patterns of inequity (such as colonialism) and marginalization. The most vulnerable people are disproportionately affected by adverse impacts of climate change.
* Over three billion people live in areas that are highly vulnerable to climate change.
* Unsustainable development patterns are increasing exposure of ecosystems and people to climate hazards.

As for projected risks, there are three key findings to highlight:

* Increased warming will cause unavoidable increases in multiple climate hazards and risks to humans and ecosystems.
* The level of risks in the future will depend on the speed and scale of mitigation efforts.

* And the level of risks in the future will depend on choices we make now about adaptation and development. We can choose to reduce vulnerability and exposure.

That last point is crucial: what are the adaptation strategies that can reduce vulnerability?

Adaptation Strategies

There are numerous ways to manage climate impacts and reduce risks and vulnerability. Whether they will be implemented, though, depends on what the IPCC calls "enabling conditions," including financing and the capacity and effectiveness of governance and policymaking processes. More on that in later sections.

There is progress being made on adaptation planning and implementation, but here too the progress is unevenly distributed. And many adaptation plans emphasize near-term climate risks by tweaking existing systems. These are the most feasible options, but they can reduce chances for transformational changes. In other words, oftentimes Band-Aids are being put onto broken bones.

We can get a sense of some adaptation strategies by looking at different kinds of risks related to:

* *Land and ocean ecosystems*: coastal and flooding defenses from mangrove and wetland restoration to dikes, levees, and sea walls and sea gates; ecosystem restoration and connectivity; species relocation, etc.
* *Water and food security*: water conservation measures; desalination; drought-resistant crop development; crop shifting; social safety nets, etc.
* *Infrastructure*: building codes and materials; building air filters (e.g., in response to increased wildfire smoke exposure); urban planning; improved forecasting and emergency response; heat-resistant rail, road, and energy systems; cooling centers, etc.
* *Cross-sectoral*: human migration; and managed retreat or planned relocation and resettlement.

Migration cuts across all aspects of climate change and adaptation, which may make it the single most important factor. Though shrouded in uncertainty, the tangle of climate change with socioeconomic issues is likely to displace (internally and across borders) millions of people. Of course, with strong adaptation measures, the displacement can be drastically reduced. Nonetheless, people are already becoming involuntary migrants. So much about our goals for shared safety and human well-being and dignity will depend on our response to migration. Will we see hardened boarders, cruelty, and violence? Or will we expand our sense of the "we" to broaden our kinship with others? Our history of conquest and colonialism doesn't offer much hope, but maybe we can write a new chapter for this new epoch.

Climate Resilient Development

The IPCC defines resilience as "the capacity of social, economic, and environmental systems to cope with a hazardous event or trend or disturbance." Climate resilient development, in other words, reduces vulnerability. This is already happening—deaths from natural disasters declined drastically across the twentieth century, even as the number of weather- and climate-related disasters has increased.

It is not guaranteed, however, that that trend will continue downward as warming continues. Indeed, the compounding and cascading risks mean that: "Climate change impacts and risks are becoming increasingly complex

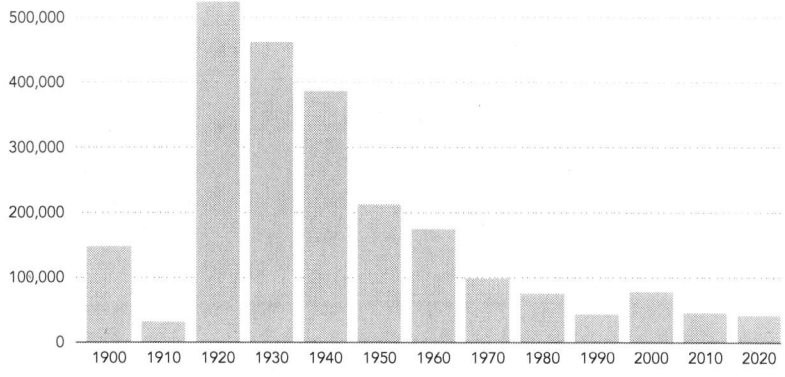

Annual Number of Deaths from Disasters, Decadal Average

Disasters include all geophysical, meteorological and climate events including earthquakes, volcanic activity, landslides, drought, wildfires, storms, and flooding. Decadal figures are measured as the annual average over the subsequent ten-year period.

FIGURE 11.7

and more difficult to manage" (IPCC 2022, p. 18). It may be that we already harvested the low-hanging fruit for adaptation in the twentieth century. Efforts will be costlier this century with each increment of warming. Dependence on past decisions constrains future decisions. For example, the design of car-centric cities makes it so much harder to build alternative forms of transport. Further, opportunities for resilient development are not equitably shared. If we take climate justice seriously, success cannot mean that the 1 percent thrive behind barricades while everyone else suffers. So, success will depend on equitable and inclusive cooperation between governments, the private sector, and civil society. In other words, it will depend on the conditions to enable wise climate action.

Finance

In Chapter 3, we encountered climate finance in the context of "common but differentiated responsibilities." We picked up on this in Chapter 9 with a discussion of loss and damage and climate reparations as a form of climate justice. This is the biggest context of climate finance—wealthy nations that are primarily responsible for climate change pledging money to aid in the resilient development of poorer nations that are disproportionately vulnerable to risks. Not nearly enough money has been pledged, let alone paid, for the tasks at hand.

In this section, we'll explore three aspects of finance understood in broad terms: a) enabling conditions and feasibility; b) insurance and investments; and c) the case of electric school buses to illustrate some of the available policy options.

Enabling Conditions and Feasibility

Mitigation and adaptation constitute enormous construction projects from solar panels and EV charging stations to dykes and retrofitted buildings. Nations, corporations, cities, and other institutions have limited resources and competing priorities. That's why it is important to analyze the conditions that can enable success in achieving climate goals. Sufficient financial resources are important but not the only enabling condition of success. There also must be political will, authority, competence, and trust in institutions. The IPCC (2022) lists several "dimensions of potential feasibility": economic, technological, institutional, social, environmental, and geophysical.

For Denton's Climate Plan, for example, home energy efficiency upgrades are one way we can bolster resilient development. However, those can be expensive and there isn't much room in the city budget to offer residents assistance. Now consider an example of a non-financial or non-economic barrier. One way to reach our net zero goal is to prohibit the installation of natural gas infrastructure in new buildings. However, the state of Texas has outlawed cities from doing that. So, absent a state-level change, we have to get creative and find a feasible route to our goal ... if one exists.

The IPCC also uses the terminology of "hard" and "soft" limits in the context of adaptation, but it also works with mitigation. I see this more as a spectrum than as an exclusive distinction. Soft limits are those that can be overcome by addressing a constraint, like limited budgets or inadequate institutional capacity. Hard limits are insurmountable. The City of Denton could allocate more money for building upgrades (soft limit) but it is largely powerless to alter state law on fuel choice (hard limit). Some species and some groups of humans have hit hard limits on their survival. The aim is to minimize that and, where possible, turn hard limits into soft limits and then overcome them—say through assisted species migration or social safety nets for people impacted by food shortages.

Investments and Insurance

Capitalism runs on investments, the commitment of resources to achieve later benefits. Companies get bank loans or sell shares on the market to investors in order to create capital for new development. The returns on that investment (if it is successful) fuel further investments and, typically, entrench economic inequalities as money flows to those who already have it.

One form of mitigation, then would be to take away the money used by coal, oil, and natural gas companies to fund extraction. This removal is known as fossil fuel divestment: institutions relinquishing their assets (e.g., stocks and bonds) connected to fossil fuel companies. Hundreds of government agencies and universities around the world have divested from fossil fuels, representing a loss of trillions of dollars for the industry (McKibben 2018).

Financial return is not the only criterion investors make when deciding where to put their money. Investment firms that manage lots of money recognize that they can also effect broader social changes. The relevant example for us is environmental, social, and governance or ESG, a framework to guide investments that includes climate change considerations and workforce diversity, equity, and inclusion. Some critique ESG for not accomplishing its stated goals and being just a form of greenwashing. As with carbon offsets, it is important to ensure that claims about climate-friendly investments are true.

Big investments need to be insured. This is especially pertinent in the case of homes and other buildings. Banks will often not write mortgages on uninsured homes and buildings. The private insurance industry provides a clear-eyed look at the risks of climate change, and it is grim. As extreme weather and disasters become more frequent and costlier, insurance rates in particularly vulnerable areas have skyrocketed. This may precipitate a crisis as people get priced out of their homes. Even more dire is the fact that some major insurance companies are pulling out of US states (like Florida, California, and Texas) with high climate risks.

The journalist Hamilton Nolan (2023) argues that there are three possible solutions:

1. *The rational capitalist solution*: Private insurers accurately price the risk and when that price is too high, people abandon places that are especially prone to climate hazards.

2. *The rational socialist solution*: Society creates programs to cushion the economic blow to help people move away from risky areas.

3. *Pirate capitalism*: This characterizes current US policy. When disaster strikes, homeowners pressure state politicians who pressure the federal government, which bails people out in a way that allows them to build in the same place in much the same way.

This last solution is unsustainable. Once private insurers abandon an area, the government can only cover the risks for so long.

Policy Options: Electric School Buses Case Study

There are too many financing policy options to offer a comprehensive account here. We'd have to consider a diverse range of products along the path from research to development to implementation. And we'd have to consider the various actors from innovators to banks to venture capitalists and the many financial instruments and policy levers at their disposal, including rebates, subsidies, low-interest government loans, and more. Further, we'd have to think through all kinds of common challenges for green financing. For example, there are principle-agent dilemmas for making energy efficient homes whenever one party (like the developer or landlord) has to pay the upfront expenses for energy savings that will benefit another party (like the future homeowner or renter).

So, instead, let's just look briefly at electric school buses, which reduce GHG emissions and have the co-benefits of lower air pollution and noise levels around children. It is a good case study, because it illustrates a central challenge in climate financing: big upfront costs that take a while to pay themselves off (see Roberts 2023). It's similar to the conversation about discount rates above—how much cost is it worth today to get a payoff tomorrow? In the long run, electric school buses save money due to lower fuel and maintenance costs. But their upfront purchase cost is two to three times as much as their diesel competitors, which is usually prohibitive for cash-strapped public school districts. So, how can we get rapid market penetration of this technology?

One idea comes from Highland Electric Fleets, which offers a new business model for bus purchases and operations. Instead of buying and maintaining the buses, school districts pay Highland a subscription fee and sign a 15-year contract. This is analogous to rooftop solar deals where the customer doesn't pay upfront for the panels, but rather just pays a monthly fee for the electricity. The Highland deal covers not just the buses, but also maintenance, replacement, charging infrastructure, training, and even bus replacement if required. This not only negates the otherwise insurmountable upfront cost, it also reduces the risks of transitioning to electric buses.

In just the first half of the 2020s, electric school buses will go from less than 2 percent of new purchases to perhaps as much as 40 percent. It is an s-curve of adoption.

There is steep growth, not just because of Highland, but also due to federal tax credits as well as federal and state grant programs. Billions of

S-curve of Adoption

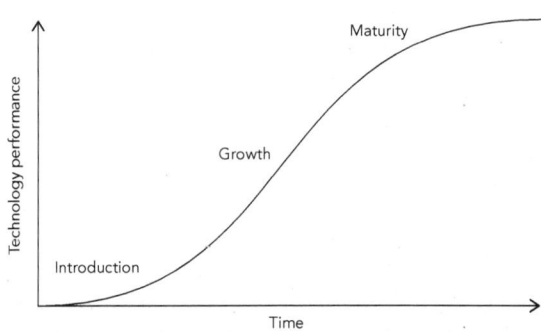

FIGURE 11.8

dollars are being poured into the effort in what the CEO of Highland called, "a tidal wave of assistance." Electric school buses show how, with creative cooperation, climate innovation can quickly transform markets and society for the better.

Conclusion

The mitigation, adaptation, and financing challenges are fascinating and complex. This chapter offered some tools to understand the landscape and direct your actions. The barriers that we face are financial and technological, to be sure, but they are foremost ethical and social. In ways that we cannot yet know, we are going to have to change our lifestyles and get accustomed to new realities. We'll have to find the will to make hard decisions in the face of uncertainty and to forge new communities in the face of clashing interests.

This is all about climate justice: Who will pay for clean energy? Who will have access to drought-resistant seeds? Whose land will be protected by dikes and who will be forced to move? Will climate change exacerbate already heartbreaking levels of inequality, or will it teach us how to relate to one another as kin, as fellow Earthlings? These are the kinds of questions we'll have to answer one way or another. That's why this book has focused on developing the moral and political skills of problem-oriented thinking. In the last chapter, we'll round things out with a look at some of the biggest debates about our role as shapers of the Earth.

Activities and Questions

1. What criteria would you use in distributing scarce resources for managed retreat projects? How would you rank the requests of different communities? What are the morally relevant considerations in such a case?

2. Research the state of the art in nuclear, biomass, and/or geothermal power. What are the debates about their feasibility, fairness, and efficacy? What place do you think they have as mitigation strategies in the grand energy transition?

3. Research the different kinds of hydrogen (e.g., gray, brown, blue, green, and white). What are the social and technical hurdles involved in making hydrogen a viable alternative to fossil fuels?

4. Does your city or town have a Climate Action Plan? If so, what is in it and is it being implemented? If not, does it make sense to create one? How would you get started?

5. Check out the Net Zero Tracker online. What surprises you about the plans of different countries and corporations?

References

Burgess, Matthew, et al. 2023. "Multidecadal Dynamics Project Slow 21st-Century Economic Growth and Income Convergence." *Communications Earth and Environment* 4 (220): 1–10. https://doi.org/10.1038/s43247-023-00874-7.

Flavelle, Christoper, and Tailyr Irvine. 2022. "Here's Where the U.S. Is Testing a New Response to Rising Seas." *New York Times*, November 2. https://www.nytimes.com/2022/11/02/climate/native-tribes-relocation-climate.html.

Graham, Jack. 2021. "Net-Zero Emissions Targets Adopted by One-fifth of World's Largest Companies." *Reuters*, March 23. https://www.reuters.com/article/us-global-climate-carbon-business-trfn/net-zero-emissions-targets-adopted-by-one-fifth-of-worlds-largest-companies-idUSKBN2BF2ZX.

Griffith, Saul. 2022. *An Optimist's Playbook for Our Clean Energy Future.* Cambridge, MA: MIT Press.

IPCC. 2020. *The Concept of Risk in the IPCC Sixth Assessment Report: A Summary of Cross-Working Group Discussions.* Geneva: Intergovernmental Panel on

Climate Change. https://www.ipcc.ch/site/assets/uploads/2021/02/Risk-guidance-FINAL_15Feb2021.pdf.

—. 2022. "Summary for Policymakers." In *Climate Change 2022: Impacts, Adaptation, and Vulnerability. Contribution of Working Group II to the Sixth Assessment of the Intergovernmental Panel on Climate Change.* Cambridge: Cambridge University Press.

Kara, Siddharth. 2023. *Cobalt Red: How the Blood of the Congo Powers Our Lives.* New York: St. Martin's.

McKibben, Bill. 2018. "At Last, Divestment Is Hitting the Fossil Fuel Industry Where It Hurts." *The Guardian*, December 16. https://www.theguardian.com/commentisfree/2018/dec/16/divestment-fossil-fuel-industry-trillions-dollars-investments-carbon.

Miller, David. 2020. *Solved: How the World's Great Cities Are Fixing the Climate Crisis.* Toronto: Aevo.

Nikiforuk, Andrew. 2023. "The Rising Chorus of Renewable Energy Skeptics." Resilience, April 10. https://www.resilience.org/stories/2023-04-10/the-rising-chorus-of-renewable-energy-skeptics/.

Nolan, Hamilton. 2023. "Insurance Politics at the End of the World." How Things Work, May 30. https://www.hamiltonnolan.com/p/insurance-politics-at-the-end-of.

Parkinson, Stuart, and Linsey Cottrell. 2022. "Estimating the Military's Global Greenhouse Gas Emissions." Scientists for Global Responsibility and the Conflict and Environment Observatory, November. https://ceobs.org/wp-content/uploads/2022/11/SGRCEOBS-Estimating_Global_Military_GHG_Emissions_Nov22_rev.pdf.

Pielke, Roger, Jr. 2022. "What Is Climate Change?" The Honest Broker, June 24. https://rogerpielkejr.substack.com/p/weekend-reading-2-what-is-climate.

Roberts, David. 2023. "Getting Electric School Buses in the Hands of School Districts." Volts, February 2. https://www.volts.wtf/p/getting-electric-school-buses-in#details.

Soga, Masashi, and Kevin J. Gaston. 2018. "Shifting Baseline Syndrome: Causes, Consequences, and Implications." *Frontiers in Ecology and the Environment* 16 (4): 222–30.

Tucker, Christopher Kevin. 2019. *A Planet of 3 Billion.* Alexandria, VA: Atlas Observatory Press.

Wallace-Wells, David. 2022. "Beyond Catastrophe: A New Climate Reality Is Coming into View." *New York Times*, October 26. https://www.nytimes.com/interactive/2022/10/26/magazine/climate-change-warming-world.html.

White, Natasha, and Akshat Rathi. 2022. "Green Groups Want Offsets Disclosed as Part of SEC's Climate Rules." Bloomberg, February 14. https://www.bloomberg.com/news/articles/2022-02-14/green-groups-want-offsets-disclosed-as-part-of-sec-s-climate-rule#xj4y7vzkg.

CHAPTER 12

Geoengineering and Rewilding

Per Capita CO$_2$ Emissions, Select Scenarios

This is measured as the global average.

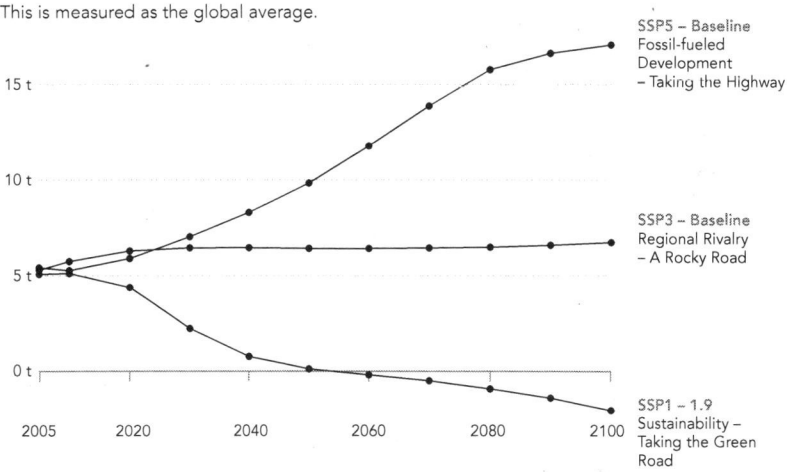

FIGURE 12.1

On March 8, 2023, Denmark's Crown Prince Frederik gave the symbolic order to begin Project Greensand. Located on a drilling platform about 200 kilometers west of the Danish coast in the North Sea, the project will store CO$_2$ gas in a sandstone reservoir over a mile below the seabed. The goal is to pump roughly eight million tons of CO$_2$ annually (equivalent to 13 percent of Denmark's CO$_2$ emissions) into the depleted reservoir of the Nini West oil field. At the ceremony, Prince Frederick underlined the historic significance of the project, which comes 50 years after his father, Prince Henrik, initiated oil and gas exploration and extraction off the Danish coast. Prince Frederick said, "It gives me great pleasure today to be able to reverse the traffic in the pipes and send CO$_2$ back into the Danish underground to the benefit of the climate for Denmark, for Europe, but also for the planet" (in Rushton 2023).

Project Greensand is the perfect symbol of our moment in history. For nearly 200 years, industrial civilization has been extracting carbon from the lithosphere and moving it, via combustion, into the atmosphere. The climatic changes this has caused mean that it is time now not only to adapt and to mitigate emissions, but also to actively reverse the flow by pulling carbon from the atmosphere and storing it in geological, land, or ocean reservoirs, or in products (like concrete or wood for construction). This is known as carbon dioxide removal (CDR), carbon capture and sequestration (or storage) (CCS), or carbon capture use and storage (CCUS), depending on the project.

The IPCC considers CCS a necessary component of reaching the net zero goal. Figure 12.1 shows one example of what it might look like to actually go net negative. In scenario SSP 1–1.9, emissions drop below zero, hinting at a future where we pull more CO_2 out of the atmosphere than we emit into it. Our present techno-politco-economic landscape is a very long way from that possible future. It is far from clear that we can develop and deploy CCS on the scales needed to get to net zero or to stay under 2°C of warming. And yet, with each passing year, the international community relies more and more on CCS.

This is impacting the way in which the IPCC talks about climate goals. There have long been debates about "phasing out fossil fuels" to reach net zero. But starting in 2022, this discourse began to shift toward phasing out *unabated* fossil fuels, meaning fossil fuels that are "produced and used without interventions that substantially reduce the amount of greenhouse gases" (see Lempriere and Evans 2023). In other words, CCS is creating a new category of energy called "abated fossil fuels." Some groups—especially the fossil fuel industry—would like to steer us toward a future that still relies on coal, oil, and natural gas but that captures and sequesters the resulting emissions. As we will see, others think that this is a very dangerous vision.

As with so many aspects of climate change, CCS is an enormous engineering challenge, beset by uncertainties, technical hurdles, political pitfalls, and ethical dilemmas. Project Greensand involves 23 international organizations in the value chain at various points from capturing and purifying the CO_2 to liquifying and transporting it to storing and monitoring it deep underground. As one example of the technical aspects involved, the CO_2 has to be compressed into a supercritical fluid[1] and injected into the right kind

---

1   The state reached by a liquid or gas under conditions of high pressure and/or temperature. In this state, the substance is neither liquid nor gas, but in some ways like each.

of reservoir rock underneath a suitable "seal" rock (e.g., shale) that will keep the CO_2 from escaping back into the atmosphere.

Other crucial complexities pertain to financing. It takes a lot of energy to move carbon from the atmosphere into the biosphere or lithosphere. Who pays for that and how? Project Greensand was helped out in this regard by the rising prices of CO_2 on the European Union Emissions Trading System. Just a month before the project launched, carbon prices hit 100 euros/ton for the first time in history. Ursula von der Leyen, the President of the European Commission, touted the project as an example of green growth, framing it in terms of innovation and the new Green Deal Industrial Plan (Brooks and Jordans 2023).

In this final chapter, we'll open a window onto aspects of climate change that we haven't had a chance to explore. We'll round out our discussion of climate policy options with a look at geoengineering. The first section builds from Project Greensand to survey a broad range of CCS types, each with different pros and cons. The next section examines various types of solar geoengineering and the controversies surrounding them. The final section turns to rewilding as a way to call our attention back to the big picture that we started with in Chapter 1.

Carbon Capture and Storage

Geoengineering refers to a broad range of techniques to directly manipulate the climate system in order to offset some of the impacts of climate change. Though "geoengineering" is the term that has stuck, it is imprecise. As the climate fiction writer Kim Stanley Robinson put it, women's empowerment can be geoengineering, because it drops the population growth rate, which impacts the biosphere, which is part of the climate system. He writes, that geoengineering is "anything that is done at scale"[2] (in Baker 2021). Given the enormous powers unleashed during the Anthropocene, we do an awful lot at scale.

Perhaps what makes the geoengineering strategies surveyed here different, then, is that they are done *directly* to the climate system in order to *intentionally* manipulate it. This is in the spirit of Stewart Brand's 1968 statement in the first issue of the *Whole Earth Catalog*: "We are as Gods and

---

2   On a large scale.

might as well get good at it." If we are shaping the entire climate system, let's do it in ways that are thoughtful and just. That's one way to think of it at any rate. In that light, "climate restoration" is a term I like better. I also think "climate intervention strategies" is pretty accurate.

Figure 12.2 puts these strategies in the context of mitigation and adaptation (explored in the last chapter). Mitigation activities aim to prevent human impacts on the climate system, and adaptation activities aim to reduce the impacts of the climate system on humans (and other species). Geoengineering strategies aim to directly intervene in the climate system.

Mitigation and Adaptation

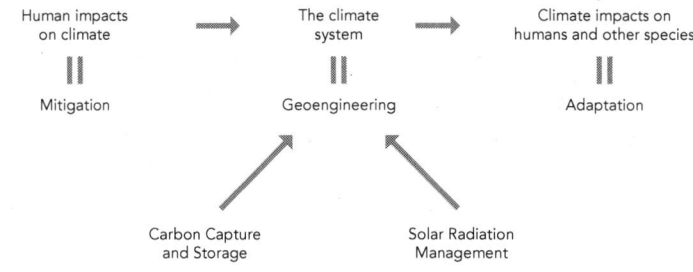

FIGURE 12.2

As this figure notes, there are two main kinds of geoengineering strategies: CCS (this section) and solar radiation management (next section). I have broken this section down into three parts: some big picture comments about CCS; a brief survey of some of the main kinds of CCS; and a look at some of their risks and downsides.

CCS in Context

The history of carbon capture or carbon dioxide removal has two main roots. The first springs from the private sector, namely the oil and gas industry. Since the 1970s, drillers have used a variety of techniques known as enhanced oil recovery (EOR) to maximize well productivity. Most of these activities involve injecting supercritical CO_2 deep underground to stimulate the flow of oil. This is much the same process utilized by Project Greensand, which is why that and similar projects rely heavily on the infrastructure and expertise of the oil and gas industry.

The second historical root of CCS is the ecological sciences that emerged in the latter half of the twentieth century. By the 1990s, several ecological specializations were studying the carbon cycle, including the Global Carbon Project and the US Carbon Cycle Science federal research program (see Shrestha 2022). By the early 2000s, these research communities were discussing "negative emissions," or the idea of removing more carbon dioxide and methane from the atmosphere than are being added to it.

After all, CCS techniques are really just ways to manipulate the carbon cycle, especially by increasing carbon sinks and the rate at which those sinks absorb carbon from the atmosphere. This interdisciplinary research project culminated in the 2015 National Academies of Sciences, Engineering, and Medicine report *Climate Intervention: Carbon Dioxide Removal and Reliable Sequestration* (NAS 2015a) as well as another report in 2019 laying out a research agenda for negative emissions technologies and sequestration (NAS 2019). By the 2020s, CDR and CCS had become massive international research programs, including over 600 US federal research projects. One example is the US Department of Energy's Carbon Negative Shot initiative, which aims to develop techniques to remove carbon at gigaton scales for less than $100/net metric ton.

In a 2018 special report, the IPCC underscored the crucial role of CCS for achieving our climate goals: "All pathways that limit global warming to 1.5°C with limited or no overshoot project the use of carbon dioxide removal (CDR) on the order of 100–1000 $GtCO_2$,[3] over the 21st century" (IPCC 2018, 17). (For some comparison, humans have emitted over 2,000 $GtCO_2$ since the industrial revolution.) In short: there is no pathway to our goals without CCS.

That broad range of the IPCC estimate represents uncertainty about many things, especially global emissions mitigation (more rapid mitigation means less need for CCS). To be clear, the bottom figure of this range is 100 billion tons. When it was launched, Project Greensand was the largest CCS project of its kind, and its goal is just 8 million tons per year, which is about 0.2 percent of the minimum need projected by the IPCC. So, there is an enormous shortfall when it comes to CCS and our temperature targets for climate policy.

---

3   Gigatonnes (billion tonnes) of carbon dioxide. A gigatonne is roughly the combined weight of all the land mammals (excluding humans) in the world.

Types of CCS

One common way to picture CCS is the bathtub analogy, where emissions flow from the tap into the atmosphere tub. Mitigation represents turning off the flow from the tap. CCS represents the drain, which has to match or exceed the rate of the flow from the tap (emissions) in order to keep the tub from overflowing, which would represent overshooting our climate budget or temperature targets.

Climate Bathtub Simulation

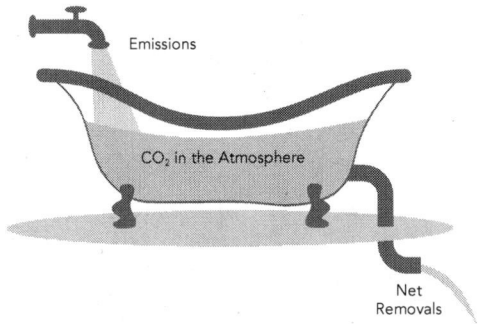

FIGURE 12.3

There are many kinds of drains or sinks in the climate system that might be manipulated to enhance or accelerate their uptake of carbon from the atmosphere. In 2023, an international consortium of scientific agencies created a report titled *The State of Carbon Dioxide Removal* (Smith et al. 2023). They use CDR in the same way I use CCS as "Capturing CO_2 from the atmosphere and storing it away for decades to millennia." This is in contrast to projects that might capture CO_2 only to use it in a fuel that is then burned, sending those emissions right back into the atmosphere. In that report, they estimate that currently there is about 2 $GtCO_2$ removal per year, a figure that needs to scale up to 5 to 10 $GtCO_2$ per year by 2050.

They group CCS techniques into two broad categories: conventional and novel. Conventional CCS (what others often call "natural") accounts for 99 percent of all CCS projects. The vast majority of these are land-based methods of afforestation, reforestation, and managing current forests, especially through the prevention of deforestation. These conventional CCS techniques are involved in the (often dubious) carbon offset practices discussed in the last chapter. Conventional CCS also includes soil carbon storage in croplands and grasslands as well as peatland and wetland restoration. CO_2 stored in the oceans and coastal ecosystems is called "blue carbon." Most nations plan to maintain or slightly increase these conventional (or natural) methods.

Nearly all scenarios to achieve climate goals call for novel types of CCS to be scaled up enormously—by a factor of at least 1,300—by 2050. Here are four of the most prominent novel techniques:

- Bioenergy with Carbon Capture and Storage (BECCS) refers to any energy pathway where CO_2 is captured from a biogenic[4] source and permanently stored in the lithosphere. For example, an ethanol fuel plant (using corn) can capture the CO_2 emissions of that process and then inject them for storage underground.

- Biochar, which is the fastest growing area of research in CCS, is black carbon produced from biomass sources through chemical and/ or thermal conversions. These conversions transform the biomass carbon into a more stable form that can be stored in soils. Indeed, this often has the co-benefit of improving soil fertility and plant productivity. A similar technique was practiced by Indigenous peoples in the Terra Preta (Dark Earth) region of the Amazon.

- Direct Air Carbon Capture and Storage (DACCS) technologies use chemical solvents to capture and separate CO_2 directly from ambient air. Orca, the world's first large-scale DACCS operation, was launched in Iceland in 2021. It captures up to 4,000 tons of CO_2 annually, storing it in the lithosphere.

- Point source capture (PSC) can utilize similar techniques to scrub CO_2 directly from large industrial facilities such as refineries or fossil-fueled electricity generating stations. In this way, the CO_2 created by combustion never reaches the atmosphere, because it is captured in the emissions pipe and diverted to the lithosphere for storage. One project in China will capture emissions from a refinery and inject them into nearby oil wells to enhance crude oil recovery. The CO_2

---

4   Something produced by a living organism.

being stored by Project Greensand was captured at an industrial facility.

There are other novel CCS techniques as well, including enhanced rock weathering, ocean iron fertilization, and ocean alkanisation.[5]

Critiques and Risks of CCS

CCS is a controversial topic among climate researchers, activists, and policy experts. Some of the downsides pertain to the specifics of each CCS technique. When it comes to conventional techniques, we already discussed the potential pitfalls of carbon credits, where industries or nations might claim carbon neutrality in less-than-honest ways. Greenwashing is a serious problem. There is also just a finite storage capacity for forests and other ecosystems, and wildfire always threatens to release the stored carbon back into the atmosphere. As with carbon offsets, CCS presents MRV (monitoring, reporting, and verification) challenges. These differ with each type, with many conventional techniques (like afforestation) being the hardest for MRV and DACCS being the easiest.

Each technique has its own mix of risks, costs, and benefits. DACCS, for example, may have the most mitigation potential (it can scale up to capture the most carbon), but it is very energy intensive, which may lead to further fossil fuel use and/or competition for other uses of renewable energies. BECCS can be land intensive, leading to competition with food, water, and other resource uses. And depending on how the biomass is harvested, biochar can have negative impacts on biodiversity and ecosystem health. Another key risk to consider for all techniques has to do with the durability of storage and its vulnerability to reversal. For example, scientists are trying to characterize the chances of CO_2 leaking from different geological reservoirs. Clearly, a massive and sudden failure of a carbon storage reservoir would be bad, but so too would be slower leaks, especially if they went undetected.

There is a bigger controversy, however, about CCS in general. Will it delay mitigation and foster a false sense of security? It could be that the urgency of acting now is deflated by the promise of technofixes later. Why

---

5   The process of making something more alkaline by addition of an alkanising substance (for example, lime).

mitigate emissions if we can just remove them? This is a very perilous thought to entertain. It is important to emphasize just how novel most CCS techniques are—relying on the uncertain promise of nascent technologies seems unwise. Some critics argue that it distracts from the source of the problem, and that all attention needs to be paid on quickly ending fossil fuel extraction (Buck 2021).

The controversy runs through the climate science community, with some seeing CCS as a dangerous distraction, and arguing that it is so unproven that we should not even consider it as an option. However, Sir David King, the former UK government chief scientific adviser, disagrees: "The carbon we have put up [in the atmosphere] will have to be removed. It may cost a fortune, but we have to recognise that the alternative is to lose our civilization" (in Harvey 2023). The key will be ensuring that carbon removal does not serve as an excuse to slow down on mitigation.

Solar Radiation Management

Capturing and storing carbon is not the only way to intervene in the climate system with the aim of cooling the planet. It is also possible to counteract temperature rise through a range of techniques that would reflect sunlight away from the Earth's surface or reduce the trapping of outgoing thermal radiation. These techniques collectively go by the name solar geoengineering, albedo modification (albedo is the proportion of light that is reflected by a surface), or solar radiation management (SRM). I use SRM here as a broad umbrella term. In this section, we'll put SRM in historical and social context, look at some of the main types of SRM, and survey some of the key arguments for and against it.

SRM in Context

SRM techniques are generally newer and less well-understood than most CCS techniques. A 2015 US National Academies report concluded that there was too much uncertainty about SRM and that it should not be deployed (NAS 2015b). Similarly, in 2018, the IPCC did not include SRM in its assessed pathways to the 1.5°C goal, concluding that the knowledge gaps were too large (IPCC 2018).

However, as the impacts of climate change have mounted and mitigation and adaptation efforts have been slower than needed, many experts are looking at SRM as a potential climate policy option. Indeed, in 2021, the National Academies returned to SRM with more urgency in a report on research and research governance in order to understand the feasibility, risks, and benefits of "the full range of options for dealing with the climate crisis" (NAS 2021).

A look at that report can give some indications of the complexities involved in trying to adjust the Earth's temperature in this way. It notes that reflecting roughly 1 percent of the sunlight the Earth absorbs may be enough to counteract all of the warming caused by anthropogenic GHG emissions. That sounds promising, but solar climate interventions and GHG emissions are different kinds of "climate forcing" mechanisms that impact the climate in very different ways. SRM can lead to various regional-scale temperature, precipitation, and other changes that are poorly understood. Climate models generally are not good at factoring in SRM kinds of climate forcings. What if some SRM intervention, for example, disrupts the monsoon cycles in India? Who would be responsible for the harms that would cause?

Types of SRM

Like CCS, there are many types of solar geoengineering. Figure 12.4 illustrates some of the most commonly researched techniques. Let's look briefly at three of them.

- Stratospheric Aerosol Injection (SAI) increases the number of small reflective particles (aerosols) in the stratosphere (an upper layer of the atmosphere) in order to increase the reflection of incoming sunlight. The most commonly studied strategy would use sulfates in a way that mimics volcanic eruptions. It is well-understood that sulfates from volcanoes have a global cooling effect. Indeed, anthropogenic aerosol pollution (primarily sulfates from SO_2 emissions) actually masks at least 0.5°C of global warming. This means that as we adopt cleaner technologies and aerosol pollution declines, warming will increase, further adding to the urgency of mitigation and geoengineering strategies. SAI could be conducted by fleets of aircraft releasing sulfates at very high altitudes. There are large uncertainties about

Solar Climate Intervention Methods

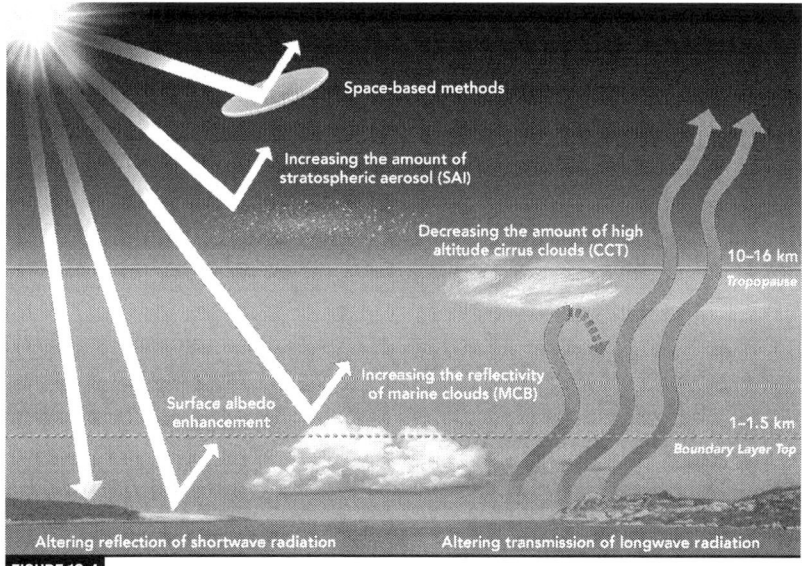

FIGURE 12.4

the injection type and amount, atmospheric chemistry interactions, and regional impacts.

* Marine Cloud Brightening (MCB) adds particles to the lower atmosphere to increase the reflectivity of low-lying clouds over the ocean. This phenomenon is commonly observed in studies of ship tracks, but there are large uncertainties about aerosol-cloud interactions and how much cloud albedo (the proportion of light reflected) can be modified and what feedbacks this might trigger. This occurs on scales too small to be included in most global climate models.

* Cirrus Cloud Thinning (CCT) modifies the properties of high-altitude ice clouds, increasing the atmosphere's transparency to outgoing thermal radiation. Cirrus clouds, on balance, warm the planet. Indeed, their warming effect is estimated to be very large, as they trap heat that would otherwise escape out of the atmosphere. In theory, drones could deliver particles to seed the clouds in ways that would make them dissipate. The efficacy of this is unknown due to limited understanding of the microphysics of cirrus clouds. The few

climate model studies of CCT thus far have yielded contradictory results due to such uncertainties.

There are other solar geoengineering techniques too. A more regional approach would be high-albedo crops and buildings. Maybe we can paint infrastructure white and introduce genes into crops to give them a reflective sheen. Space sunshades is another idea. We could deploy giant mirrors in outer space, which could reflect a small portion of sunlight that would otherwise hit the Earth. This would perhaps be the least environmentally disruptive technique, but also likely the costliest.

Critiques and Risks of SRM

As with the CCS techniques, each kind of SRM has its own unique profile of feasibility, risks, and benefits. More generally, I think there are two overarching criticisms of SRM:

- *Hubris*: It is arrogant for us to think we could tune the temperature of the Earth like setting the thermostat in a house. The climate is a complex, chaotic, dynamic system, and we are bound to trigger unintended consequences that may be catastrophic, at least for certain ecosystems and parts of the human population.

- *Distraction and delay*: As with CCS, there is a real concern that even researching SRM might lead to complacency and a false sense of security that will drain the urgency of mitigation and perhaps even justify further fossil fuel extraction. Why worry so much about GHG emissions when we can use CCS to store carbon or SRM to cool things down? This kind of concern is often called a "moral hazard," a lack of incentive to guard against risks due to a feeling of being protected from their consequences.

This is why the 2021 NAS report insists that SRM is not a substitute for mitigation, but a stopgap measure to deal with near term climate risks. They list four reasons for caution about SRM:

1. It does not address the underlying driver of climate change (GHG emissions), and it does not address other impacts of climate change like ocean acidification.

2. It raises concerns about new risks and unintended impacts on natural ecosystems, agriculture, and human health and safety.

3. It does not provide a reliable means to restore climate to some desired prior state; and

4. It entails the risk of catastrophic, rapid warming if the intervention were halted prior to suitable levels of mitigation. This is known as "termination shock."

For such reasons, the NAS report recommends a research project that is careful, coordinated, and transparent and that includes a diverse range of disciplines and stakeholders.

Rewilding

For the last section of the book, it is fitting to recall the big picture we began with in Chapter 1. Whether we call it the Anthropocene, the Capitalocene, the Chthulucene (Haraway 2016), or the Great Acceleration, it is clear that climate change is part of massive, rapid global changes across many dimensions. Earlier, we termed this a "polycrisis," or we might just say a tangle of crises. As we have seen, most international policy discourse around climate change is focused on GHG emissions. There is good reason for this, but it can also limit our vision of the bigger picture. We should beware "carbon tunnel vision" where we focus solely on emissions, as if that alone will bring us to a fair and flourishing future. We have to see all the strands in the tangle of problems.

To me, the most important thing not to lose sight of is biodiversity loss, or better, *biocultural* diversity loss. The development project that is sweeping the planet tends to homogenize ecosystems and cultures. This is deeply shaped by the legacy of colonialism and the logic of capitalism. It is what the conservationist Aldo Leopold (1949) called "ecological murder." I think of rewilding broadly as efforts to reverse this global trend of losing the mar-

Carbon Tunnel Vision

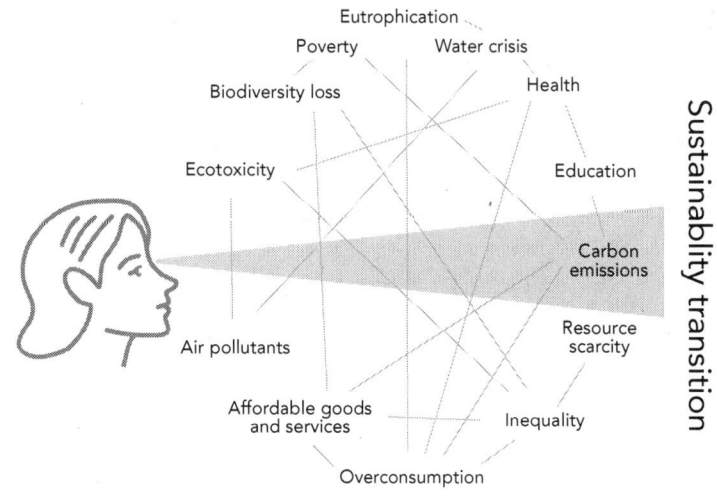

FIGURE 12.5

velous, beautiful diversity of Earth. This includes ecosystem diversity—wild places—along with species diversity. Can we imagine a future of biodiversity—and biocultural diversity—*increase*? How would we get there?

This is an area of intense research and debate. One of the most ambitious and influential proposals comes from the renowned biologist E.O. Wilson in his book *Half-Earth* (2017). He argues that the only way to stave off a disastrous collapse is to devote half the surface of the Earth to nature. In terms of international biodiversity politics, the United Nations and its Sustainable Development Goals play important roles. For example, at the UN Biodiversity Conference in 2022, 188 nations agreed to the Kunming-Montreal Global Biodiversity Framework, which aims to address biodiversity loss and protect Indigenous rights. It outlines 23 targets, including putting 30 percent of the planet under protection by 2030.

Rewilding is similar to ecological restoration, but there are a couple of key differences. First, rewilding efforts generally try to minimize human influence on ecosystems, preferring instead to "let nature take her course" in ways that create self-sustaining ecosystems. By contrast, restoration projects tend to be more heavily reliant on human interference. Second, restoration aims at recreating ecosystems that existed prior to human interference. This can be a goal of rewilding too, but it is also open to novel or emerging ecosystems that may have new species and new interspecies relationships.

It is most fruitful here to blur the human-nature dichotomy and think in terms of interspecies politics. For example, restoration has a lot of resonance with reparations, which the *Oxford English Dictionary* defines as "to make amends for; to compensate or make good (loss or damage)." Leopold was adamant on this point as he wrote from his experiences watching in horror as diverse midwestern prairie ecosystems were converted into monocultures of corn and soybeans. Those crop yields are so high because of the millennia of work put in by prairie plants to build the deep, fertile topsoil. Don't those of us who eat diets dominated by corn and soybeans owe something to the prairie? We are drawing down the fund of fertility that prairie species created without contributing back to that fund. Maybe some efforts to store carbon in soils could be framed in terms of prairie reparations.

It may also be useful to recall the rocket ship and footprint metaphors discussed in Chapter 4 on green growth and degrowth. It seems clear that any practical rewilding effort will need a blend of both paradigms: innovation as well as cultural adjustments (see Lewis 2015). Technological and economic developments can reduce the need to extract resources, which can free up more areas for rewilding. Indeed, improved crop yields and the transition from wood to coal and natural gas for energy have already led to rewilding efforts in some areas that are no longer used for agriculture and resource extraction. Sometimes rewilding even happens spontaneously in such cases. As improvements to vertical agriculture and other techniques accrue, it is possible that even more land will become available for rewilding. Of course, whether or how rewilding comes about will be a political, and not solely a technical, matter.

As with the reintroduction of wolves in the greater Yellowstone ecosystem, rewilding often takes lots of planning and policy. In that case, the gray wolf had been extirpated from the area by the early twentieth century (the last wolves were killed in 1926). Beginning in the 1940s, park managers and conservationists started working on plans to reintroduce wolves. Yet it would take the 1973 passage of the Endangered Species Act and over 20 more years of work. It wasn't until 1995 that gray wolves were reintroduced into Yellowstone in the Lamar Valley. So, as this case shows, innovation and market forces on their own certainly won't usher in a "Good Anthropocene" of flourishing human and natural systems. We will need a mix of concerted conservation and policy efforts to save terrestrial and aquatic ecosystems with breakthroughs in technology that can also reduce the human burden on other species.

Fascinating and profound political and ethical debates lie ahead of us on this aspect of climate change. Some of these pertain to what we will consider "wild," given that even wilderness areas are touched by human influence and require some forms of management (e.g., to extirpate invasive species, a concept that is also up for debate). Other debates will form around specific policy proposals. For example, could precise fermentation end our reliance on agriculture as we know it, perhaps freeing up millions of square kilometers of land? Should we revive the wooly mammoth and seek to develop viable populations in contemporary ecosystems?

Conclusion

In order to achieve our goal of climate safety, we are going to need to pull carbon out of the atmosphere and sequester it in the ground. How much carbon we will have to store depends on how rapidly the global economy decarbonizes through mitigation efforts. It is less clear whether or to what extent solar geoengineering can or should be part of climate policy in the future. Ideas like seeding the stratosphere with sulphates or deploying mirrors in space seem like science fiction, but we know that climate change also entails changes to our sense of what is ordinary or normal. Rewilding speaks to our larger aspirations not just for safety, but for flourishing. After all, I think we all want a future of abundance and biocultural diversity. To get there, we'll have to work through a tangle of problems. I hope this book has helped in that journey.

Activities and Questions

1. Geoengineering seems to be another paradox—what *is* it: a dangerous distraction or a vital tool for achieving our climate goals? Which, if any, of the techniques do you think should be further studied or implemented and why?

2. Imagine you are in charge of designing an international solar radiation management treaty: who would be involved and how would decisions

be made? What would a fair process and a good outcome look like? What would your rules be for deliberate outdoor experiments in SRM?

3. What does "wild" mean and what is "wilderness"? What makes a species "native" or "invasive," and what are the political implications of these categorizations? Research horses in North America. Are they native? Should they be removed or restored? Who should decide that and by what criteria and norms?

4. Research the concept of energy returned on energy invested (EROEI) or energy return on investment (EROI). How might this apply to carbon capture and other aspects of climate policy?

References

Baker, Aryn. 2021. "Kim Stanly Robinson on Climate Change." *Time*, September 8. https://time.com/6086585/kim-stanley-robinson-climate-change-fiction/.

Brooks, James, and Frank Jordans. 2023. "Denmark Hopes to Pump Some Climate Gas beneath the Sea Floor." *AP News*, March 8. https://apnews.com/article/denmark-offshore-carbon-capture-od18eeba095b3584c8b4390f51570905.

Buck, Holly Jean. 2021. *Ending Fossil Fuels: Why Net Zero Is Not Enough*. New York: Verso.

Haraway, Donna. 2016. *Staying with the Trouble: Making Kin in the Chthulucene*. Durham, NC: Duke University Press.

Harvey, Fiona. 2023. "Carbon Dioxide Removal: The Tech that Is Polarising Climate Science." *The Guardian*, April 25. https://www.theguardian.com/environment/2023/apr/25/carbon-dioxide-removal-tech-polarising-climate-science.

IPCC. 2018. "Summary for Policymakers." In *Global Warming of 1.5°C: An IPCC Special Report*, 3–24. Cambridge: Cambridge University Press.

Leopold, Aldo. 1949. *A Sand County Almanac: And Sketches Here and There*. Oxford: Oxford University Press.

Lewis, Martin. 2015. "Rewilding Pragmatism." The Breakthrough Institute, June 3. https://thebreakthrough.org/journal/issue-5/rewilding-pragmatism.

NAS. 2015a. *Climate Intervention: Carbon Dioxide Removal and Reliable Sequestration*. Washington, DC: The National Academies Press.

—. 2015b. *Climate Intervention: Reflecting Sunlight to Cool the Earth*. Washington, DC: The National Academies Press.

—. 2019. *Negative Emissions Technologies and Reliable Sequestration.* Washington, DC: The National Academies Press.

—. 2021. *Reflecting Sunlight: Recommendations for Solar Geoengineering Research and Research Governance.* Washington, DC: The National Academies Press.

Rushton, Simon. 2023. "Denmark Injects Carbon Dioxide into Undersea Storage in World First." *The National News*, March 8. https://www.thenationalnews. com/world/europe/2023/03/08/denmark-injects-carbon-dioxide-into-undersea-storage-in-world-first/.

Shrestha, Gyami. 2022. "What Is the Big Picture in Carbon Removal Research?" *Issues in Science and Technology* 39, no. 1 (Fall): 17–19. https://issues.org/carbon-dioxide-removal-research-coordination-gyami-shrestha/.

Smith, Stephen, et al. 2023. *The State of Carbon Dioxide Removal: A Global, Independent Scientific Assessment of Carbon Dioxide Removal.* https://www. stateofcdr.org/resources.

Wilson, E.O. 2017. *Half-Earth: Our Planet's Fight for Life.* New York: Liveright.

Permissions Acknowledgments

0.2 and 0.3 Global Warming's Six Americas, Fall 2023, Yale Program on Climate Change Communication; George Mason University Center for Climate Change Communication. Reprinted with permission.

1.1 A Ball-and-Cup Depiction of the Earth System Anthropocene. Fig. 4 from W. Steffen, et al., "Stratigraphic and Earth System Approaches to Defining the Anthropocene," *Earth's Future* 4, no. 8 (August 2016): 324–45, https://doi.org/10.1002/2016EF000379. Reprinted by permission of John Wiley & Sons, Inc., conveyed through Copyright Clearance Center, Inc.

2.1 Yearly Global Surface Temperature and Atmospheric Carbon Dioxide (1850–2022). https://www.climate.gov/media/13840. Original graph by Dr. Howard Diamond (NOAA ARL), and adapted by NOAA Climate.gov. Reproduced with adaptations.

2.2 The Climate System. Fig. 1.2, "Earth's Climate System," illustrated by Walther-Maria Scheid, from *WOR 1: Living with the Oceans*, 2010, World Ocean Review published by maribus gGmbH. Reprinted by permission of the illustrator.

4.5 Nine Metrics for Human Well-Being and Per Capita Energy Consumption. Fig. 1, "Human Well-Being and Per Capita Energy Use," from R.B. Jackson, et al., *Ecosphere* 13, no. 4 (April 2022): e3978, https://doi.org/10.1002/ecs2.3978. Used under CC BY Attribution 4.0 International license, https://creativecommons.org/licenses/by/4.0/.

5.3 IPCC Organization Chart, IPCC 2023: Structure of the IPCC. https://www.ipcc.ch/about/structure/. Reprinted by permission of The Intergovernmental Panel on Climate Change.

6.1 Global Cumulative Glacier Ice Loss. Fig. 2, "Global Glacier State," World Glacier Monitoring Service, https://wgms.ch/global-glacier-state/.

6.3 History of Global Temperature Change and Causes of Recent Warming. Fig. SPM.1 from *IPCC, 2021: Summary for Policymakers*, from *Climate Change 2021: The Physical Science Basis. Contribution of Working Group I to the Sixth Assessment Report of the Intergovernmental Panel on Climate Change* (Masson-Delmotte, V., P. Zhai, A. Pirani, S.L. Connors, C. Péan, S. Berger, N. Caud,

Y. Chen, L. Goldfarb, M.I. Gomis, M. Huang, K. Leitzell, E. Lonnoy, J.B.R. Matthews, T.K. Maycock, T. Waterfield, O. Yelekçi, R. Yu, and B. Zhou, eds.). Cambridge University Press, Cambridge, UK and New York, NY, USA, pp. 3–32, doi:10.1017/9781009157896.001. Used by permission of the IPCC. Reproduced with adaptations.

6.4 Process for Evaluating and Communicating the Degree of Certainty in Key Findings. Fig. 1 from M.D. Mastrandrea, K.J. Mach, G-K. Plattner, et al., "The IPCC AR5 Guidance Note on Consistent Treatment of Uncertainties: A Common Approach across the Working Groups," *Climatic Change* 108 (2011): 675–91, https://doi.org/10.1007/s10584-011-0178-6. Used by permission of the IPCC. Reproduced with adaptations.

6.5 A Depiction of Evidence and Agreement Statements and Their Relationship to Confidence. Fig. 2 from M.D. Mastrandrea, K.J. Mach, G-K. Plattner, et al., "The IPCC AR5 Guidance Note on Consistent Treatment of Uncertainties: A Common Approach across the Working Groups," *Climatic Change* 108 (2011): 675–91, https://doi.org/10.1007/s10584-011-0178-6. Used by permission of the IPCC.

7.2 Global Observing System, World Meteorological Organization (WMO). Used with permission.

7.3 ECHO-Rover Approaches an Emperor Penguin Colony in Atka Bay, Antarctica. Image courtesy of Aymeric Houstin, FAU & WHOI.

7.4 The Evolution of Climate Models. Infographic by Rosamund Pearce, based on the work of Dr. Gavin Schmidt, Carbon Brief, https://www.carbonbrief.org/qa-how-do-climate-models-work/. Used with permission.

7.5 Global CO_2 emissions ($GtCO_2$) for All IAM Runs in the SSP Database Separated Out by SSP. Chart courtesy of Glen Peters and Robbie Andrews and the Global Carbon Project, https://www.carbonbrief.org/explainer-how-shared-socioeconomic-pathways-explore-future-climate-change/. Reproduced with adaptations.

10.1 The Colorado River Basin, United States Geological Survey (USGS).

11.1 Simplified Emissions Pathways for Climate Targets. Courtesy of Zeke Hausfather.

11.5 Estimated US Energy Consumption in 2022. Lawrence Livermore National Laboratory, July 2023. Data based on DOE/EIA SEDS (2021). Courtesy of Lawrence Livermore National Laboratory. Reproduced with adaptations.

12.3 Climate Bathtub Simulation. Reproduced by permission of Dr. John Sterman and Climate Interactive.

12.4 Solar Climate Intervention Methods. Chelsea Thompson, University of Colorado/CIRES and NOAA Chemical Sciences Laboratory.

12.5 Carbon Tunnel Vision. Reproduced by permission of Dr. Jan Konietzko.

Used under Creative Commons CC BY 4.0, https://creativecommons.org/licenses/by/4.0/

0.1 Global Annual CO_2 Emissions. Hannah Ritchie, et al., "CO_2 and Greenhouse Gas Emissions," OurWorldInData.org, 2023, https://ourworldindata.org/co2-and-greenhouse-gas-emissions. Reproduced with adaptations.

1.3 The World Population over the Past 12,000 Years. Hannah Ritchie, et al., "Population Growth," OurWorldInData.org, 2023, https://ourworldindata.org/population-growth. Reproduced with adaptations.

1.4 World GDP over the Past 2,000 Years. Max Roser, et al., "Economic Growth," OurWorldInData.org, 2023, https://ourworldindata.org/economic-growth.

1.5 Annual CO_2 Emissions by World Region. Hannah Ritchie, et al., "CO_2 and Greenhouse Gas Emissions," OurWorldInData.org, 2023, https://ourworldindata.org/co2-and-greenhouse-gas-emissions.

3.1 Global Greenhouse Gas Emissions and Warming Scenarios. Hannah Ritchie, et al., "CO_2 and Greenhouse Gas Emissions," OurWorldInData.org, 2023, https://ourworldindata.org/co2-and-greenhouse-gas-emissions. Reproduced with adaptations. Reproduced with adaptations.

3.2 Global Carbon Budget for a Two-Degree World. Hannah Ritchie and Pablo Rosado, "Fossil Fuels," OurWorldInData.org, 2017, https://ourworldindata.org/fossil-fuels. Reproduced with adaptations.

3.3 Global Deaths from Disasters over More Than a Century. Hannah Ritchie, "A Century of Global Deaths from Disasters," OurWorldInData.org, 2022, https://ourworldindata.org/century-disaster-deaths. Reproduced with adaptations (numbers added to select incidents).

3.4 GDP Per Capita, 1650 to 2018. Max Roser, et al., "Economic Growth," OurWorldInData.org, 2023, https://ourworldindata.org/economic-growth. Reproduced with adaptations.

4.1 Decoupling CO_2 Emissions and GDP in the UK. Hannah Ritchie, et al., "CO_2 and Greenhouse Gas Emissions," OurWorldInData.org, 2023, https://ourworldindata.org/co2-and-greenhouse-gas-emissions. Reproduced with adaptations.

4.2 Global Emissions of Ozone-Depleting Substances. Hannah Ritchie, et al., "Ozone Layer," OurWorldInData.org, 2023, https://ourworldindata.org/ozone-layer. Reproduced with adaptations.

5.1 Global Atmospheric CO_2 Concentrations. Hannah Ritchie, et al., "CO_2 and Greenhouse Gas Emissions," OurWorldInData.org, 2023, https://ourworldindata.org/co2-and-greenhouse-gas-emissions.

6.2 Prevalence of Undernourishment in Developing Countries, 1970–2015. Hannah Ritchie, et al., "Hunger and Undernourishment," OurWorldInData.org, 2023, https://ourworldindata.org/hunger-and-undernourishment.

9.1 CO_2 Emissions Per Capita vs. GDP. Hannah Ritchie, et al., "CO_2 and Greenhouse Gas Emissions," OurWorldInData.org, 2023, https://ourworldindata.org/co2-and-greenhouse-gas-emissions. Reproduced with adaptations (select countries shown).

9.2 Greenhouse Gas Emissions Per Kilogram of Food Product. Hannah Ritchie, et al., "Environmental Impacts of Food Production," OurWorldInData.org, 2022, https://ourworldindata.org/environmental-impacts-of-food. Reproduced with adaptations.

9.5 Fair Shares of Global Carbon Budgets. Fig. 1 from A.L. Fanning and J. Hickel, "Compensation for Atmospheric Appropriation," *Nature Sustainability* 6 (2023): 1077–86, https://doi.org/10.1038/s41893-023-01130-8.

10.4 Global Energy Consumption by Source. Hannah Ritchie, et al., "Energy," OurWorldInData.org, 2023, https://ourworldindata.org/energy. Reproduced with adaptations.

11.3 The Kaya Identity and Global Drivers of CO_2 Emissions. Hannah Ritchie, et al., "CO_2 and Greenhouse Gas Emissions," OurWorldInData.org, 2023, https://ourworldindata.org/co2-and-greenhouse-gas-emissions. Reproduced with adaptations.

11.4 Global Greenhouse Gas Emissions by Sector. Hannah Ritchie, "Sector by Sector: Where Do Global Greenhouse Gas Emissions Come From?," OurWorldInData.org, 2020, https://ourworldindata.org/ghg-emissions-by-sector. Reproduced with adaptations.

11.7 Annual Number of Deaths from Disasters, Decadal Average. Hannah Ritchie and Pablo Rosado, "Natural Disasters," OurWorldInData.org, 2022, https://ourworldindata.org/natural-disasters.

12.1 Per Capita CO_2 Emissions, Select Scenarios. Data source: K. Riahi, et al., "The Shared Socioeconomic Pathways and Their Energy, Land Use, and Greenhouse Gas Emissions Implications: An Overview," *Global Environmental Change* 42 (2017): 153–68; OurWorldInData.org, https://ourworldindata.org/explorers/ipcc-scenarios.

Used under Creative Commons CC BY-SA 4.0, https://creativecommons.org/licenses/by-sa/4.0/

4.4 Doughnut Economics. Wikimedia Commons/DoughnutEconomics.

8.1 East African Crude Oil Pipeline. Wikimedia Commons/Sputink.

Used under Creative Commons CC BY 2.0 DEED, https://creativecommons.org/licenses/by/2.0/deed.en

7.1 Havasupai Point from the Grand Scenic Divide. Photograph by Henry G. Peabody, c. 1899. Glass plate negative number 1935, Grand Canyon National Park Museum Collection. Wikimedia Commons/Grand Canyon National Park.

Index

About the Publisher

The word "broadview" expresses a good deal of the philosophy behind our company. Our focus is very much on the humanities and social sciences—especially literature, writing, and philosophy—but within these fields we are open to a broad range of academic approaches and political viewpoints. We strive in particular to produce high-quality, pedagogically useful books for higher education classrooms— anthologies, editions, sourcebooks, surveys of particular academic fields and sub-fields, and also course texts for subjects such as composition, business communication, and critical thinking. We welcome the perspectives of authors from marginalized and underrepresented groups, and we have a strong commitment to the environment. We publish English-language works and translations from many parts of the world, and our books are available world-wide; we also publish a select list of titles with a specifically Canadian emphasis.

broadview press